D0482174

WITHDRAWN

# THE LIFE OF SUPER-EARTHS

# THE LIFE OF SUPER-EARTHS

## How the Hunt for Alien Worlds and Artificial Cells Will Revolutionize Life on Our Planet

DIMITAR SASSELOV

BASIC BOOKS
A Member of the Perseus Books Group
New York

Published by Basic Books,
A Member of the Perseus Books Group



Illustrations by Sandra L. Cundiff and Michael Hardesty

Designed by Trish Wilkinson
Set in 11.5 point Goudy Old Style

Library of Congress Cataloging-in-Publication Data
Sasselov, Dimitar D.
The life of super-Earths : how the hunt for alien worlds and artificial cells will revolutionize life on our planet / Dimitar Sasselov.
p. cm.
Includes bibliographical references and index.
ISBN 978-0-465-02193-2 (hardback)—ISBN 978-0-465-02340-0 (e-book) 1. Exobiology. 2. Synthetic biology. 3. Extrasolar planets. 4. Life—Origin. 5. Life on other planets. I. Title.
QH326.S27 2012
576.8'39—dc23 2011036888

10 9 8 7 6 5 4 3 2 1

# CONTENTS

## PART II
## ORIGINS OF LIFE

# ACKNOWLEDGMENTS

This book grew out of the general education lecture course, Life as a Planetary Phenomenon, that my colleague Andrew Knoll and I designed and have taught together at Harvard since 2005; I am much indebted to Andy for encouraging and teaching me along the way.

My book introduces a general audience to new ideas on old big questions about life and the cosmic perspective on life. These questions formed the foundation for the research agenda of the Harvard Origins of Life Initiative and its core team of scientists—I thank my colleagues for the amazing collaboration and for teaching me what little I know in their fields. In particular, I owe much to Jack Szostak for being a great teacher and partner from the very beginning; to Andy Knoll and my late friend Mike Lecar for getting me into this in the first place; to George Whitesides for his wise advice about chemistry and much more; to Stein Jacobsen, Scot Martin, George Church, Ann Pearson, Rick O'Connell, David Latham, and Sarah Stewart for pushing the boundaries. I was inspired by the books of Erwin Schroedinger and Freeman Dyson and by the brave experiments of Gerald Joyce and Craig Venter, and many others; I apologize to all whose contributions are not mentioned in my short book.

My goal was to write a popular book because the new science concepts are truly wonderful and exciting, and because they have immediate implications for all of us. My approach was to provide a thorough introduction in order to make the science accessible, followed by my own views on the unanswered questions, and keeping the technical details to the endnotes only. Thanks to John Brockman and Katinka Matson my book took the right path toward that goal; my agent Max Brockman made sure the project was accomplished—I owe them a lot for their guidance and support.

The road from lab to paper and to readable prose is torturous; I was incredibly lucky to have TJ Kelleher as my editor, from the first meeting years ago to the final touches. He understands the science and he is a talented writer! My friend Peter Abresch helped me with the baby steps and showed me, in my first chapter, how to write well. To both of them I am very grateful, as I am to Andy Knoll for his critical reading of the biology-related chapters of the book. The beautiful and intelligent illustrations are the work of gifted artists Sandra Cundiff and Michael Hardesty, and I am very grateful to them.

My deepest thanks are due to my family because this book would not have come to be without their encouragement, patience, and support: to my dear parents whom I owe for who I am and what I can do, and to my dear wife, Sheila, for being by my side throughout the entire process.

# INTRODUCTION

## *The Mystery of Life*

There are few big questions that rival this one: What is life and how did it come to be? It has always been a big question, though not always for science alone. And there have always been numerous models, scenarios, speculations, and ideas offered in response—most of them not terribly successful. The middle of the nineteenth century was no different. But some samples of slimy mud scooped up from the depths of the North Atlantic along the route of a telegraph cable would change that.

The year 1857 could be celebrated as the time humanity took the first practical step to create a global world on this planet—the global-awareness world we live in today. Converted British and American warships laden with rolls of cable were laying the first intercontinental telegraph connection on the bottom of the Atlantic Ocean between Europe and America. The human timescales of news traveling on foot or by horse or by pigeon were giving way, ultimately to be replaced by instantaneous communication at the speed of light. Days and weeks were being replaced by hours and minutes. It seemed as if the oceans that had separated humans for millennia "had suddenly dried up," as newspapers at the time wrote.

In preparation for laying the telegraph cable, ships like HMS *Cyclops* and USS *Arctic* were sounding the Atlantic Ocean floor and sampling the ocean bottom. In 1868 Thomas Henry Huxley, an English biologist with major achievements in comparative anatomy (although better known today for his role as a popularizer of Darwin's theory of evolution), discovered among the samples taken from the Atlantic a substance—gelatinous, colorless, and formless—that he thought was a new life-form. Not just any life-form, Huxley thought, but the primordial organic substance, the undifferentiated protoplasm from which life originates.

It was an audacious idea for a heady time in the quest to understand life and its origins, and Huxley was in the middle of it all. First, in 1859 Charles Darwin published his seminal *On the Origin of Species*, and the theory of evolution had become a topic of broad and heated debate. Then, between 1860 and 1863, Louis Pasteur completed his famous experiments with sterilization. Between them, long-held concepts about the origin of life were being completely upended.

Before Darwin and Pasteur, Western science had attempted to explain life's origins through a combination of spontaneous generation and vitalism. Spontaneous generation was the idea that life emerges from decomposing matter, the latter being imbued by a vital force (common to all organic material, and the air as well). Vitalism was already under attack from chemistry. In its early development, chemistry had separated inorganic compounds from organic ones, the latter being erroneously assumed to derive from living forms only. Once an organic compound was synthesized in a laboratory in 1828, the need for a vital force was on its way out (although organic chemistry still keeps its name).

The fallacy of spontaneous generation had been exposed in experiments involving extensive boiling of meat broths before Pasteur, but his elegant experiments allowed access to air and thus proved that life emerges only from life. The long, sharply curved swan-neck flasks that he used to boil the broth prevented germs (i.e., bacteria and spores) from entering the sterilized liquid but still let in air. It appears that Pasteur convinced everyone.

None of that could help scientists understand life's origin, except that now they could clearly state the problem. Both Pasteur and Darwin described the origin as a single act of abiogenesis: that the first life-form emerged from inanimate matter, which happened just once. For Pasteur it was an act of God's creation, while Darwin left it to a "warm little pond," according to a letter he wrote in 1871.

Against this backdrop, it is no wonder that Huxley thought he had something big on his hands. Indeed, he named the discovery *Bathybius haeckelii*, for the German biologist Ernst Haeckel, who had recently proposed that all life descended from a primordial ooze that he called *Urschleim*. Indeed, Huxley was convinced that he had found the *Urschleim*, and the "discovery" helped prompt the dispatch of the HMS *Challenger* on a systematic exploration of the depths of the world oceans. No trace of *Bathybius haeckelii*, or *Urschleim*, was found; instead, the chemist aboard the ship found that Huxley's curious substance was simply a chemical precipitate (a hydrated calcium sulfate). In 1875 Huxley acknowledged his error.

The hunt for beginnings has never ceased, despite Huxley's error. The twentieth century had its share of milestones and conceptual breakthroughs, though sometimes they felt like a replay of nineteenth-century events but at the molecular level: the germs and microbes were replaced with the molecules of life, but the mystery surrounding life remained.

In 1953 Stanley Miller, working in Harold Urey's lab, showed that amino acids—the building blocks of all proteins, and the same protein compounds Darwin mused had formed chemically in the "warm little pond"—can be synthesized in a flask containing ammonia, methane, water, and an electric discharge. Good first step! In the same year Watson and Crick resolved the structure of the DNA molecule. It was the high point for twentieth-century biology as a whole, but much less so for research into the origin of life: how could primitive life come up with such a complex molecule?

Next came a gift from the heavens, literally, with the Murchison meteorite that fell in Australia in September 1969. Quick analysis of this piece of primitive unprocessed material from the early history of the Solar System revealed a rich set of organic molecules and many amino acids among them—not that different from the ones synthesized in the Miller-Urey experiments. Here we had rocky material that had never been incorporated into a big planet or asteroid, though from a big enough chunk that had warmed up just enough to briefly have liquid water inside it, and the primitive material had produced the building blocks for proteins by pure chemistry. Studies in 2008 and 2010 have revealed about 14,000 different organic compounds, including two nucleobases.

As exciting as these discoveries are, they still don't answer our big question. The actual origin of life on Earth remains as elusive as ever and may well stay that way. After all, it is a historical question that requires knowing environments that are not preserved in the Earth's geological record. The more general question—about possible pathways from chemistry to life—appears more within reach of today's science.

Astronomy and the hunt for exoplanets—planets orbiting other stars—offer an approach to the problem. Exploring other Earth-like planets gives us the opportunity to investigate analogs of our own planet under conditions that held before life emerged. This approach has been wildly successful in astronomy. We study stars by proxy, getting to "know" our Sun through time by examining similar stars at other stages of their lifecycle. So, in a sense, we can answer the general questions about the origins of life, about what life is, and how environments determine its appearance, by simply asking, Is there life on other planets? There are more stars in the Universe than there are grains of sand in all the beaches on the

Earth. And there are at least as many planets as there are stars. If only 1 percent of them are like Earth, does this make life on them inevitable?

Astronomy has always been about big numbers—astronomical numbers—and experience with big numbers has taught us that they do not guarantee inevitability. We have to go and find out for ourselves. Still, it seems likely that on some of those Earth-like planets, we will find signs of life. When we discover New Earth—a planet we could call home—the question of the "plurality of worlds" will come front and center, reminding us yet again that we are not the center of the Universe. The Copernican revolution, which placed the Sun, not Earth, at the center of our system of planets, did it first. That shift launched modern science and technology. Today, two efforts have placed us on the verge of completing the Copernican revolution. One is the discovery of a new Earth. The other is the era of synthetic biology. These two milestones are going to teach us about our place in the universe in ways we could never have imagined.

Want a front-row seat for these unfolding events? Climb aboard and we'll get under way.

# PART I

# SUPER-EARTH

## CHAPTER ONE

# EXTRASOLAR PLANETS AT LAST

In October 1995, I was attending a conference in Florence, Italy, that beautiful old city where the Medicis were the patrons of astronomy during the seventeenth century. I was there to exchange new ideas and thinking with my colleagues. Then during an unguarded moment of casual conversation, as often happens, a bold new concept exploded amid my deeply held presumptions.

At the day's end, a couple of us were talking to Swiss astronomer Michel Mayor about his discovery of a small companion—a planet about the size of Jupiter—around a star named 51 Pegasi. The claim itself was not a "wow" moment; such claims had come and gone in decades past. What really caught my attention was that Michel and his graduate student, Didier Queloz, had measured the orbital period in days, not in years, as one would expect. This new planet circled its sun in just four hundred days!

I was incredulous.

Okay, stars are my specialty, not planets, but I know the basics, and this did not fit. As far back as my last year of high school I had known about the Kant-Laplace model of the formation of our Solar System. Although you may know Immanuel Kant as a philosopher, as a young man he was an astronomer and an Isaac Newton groupie. He was at the University of Koenigsberg, today's Kaliningrad on the Baltic Sea, and he used Newton's new calculus and theoretical mechanics to solve an obvious but unexplained feature of the Solar System.

Astronomers before Kant had noted that all planets orbit the Sun in the same plane and in the same direction, which is also the direction in which the Sun spins. Most planets spin that way as well. Kant offered an elegant solution for this by analogy with Saturn's rings. Planets form from particles circling the sun in a rotating flat disk, and the conservation of angular momentum explains its flattened shape.* (Because his publisher went bankrupt, Kant didn't get the credit due him at the time, as recounted in *The Discovery of Our Galaxy* by my old mentor, Charles Whitney.)[1] Pierre-Simon Laplace added mathematical rigor to Kant's ideas in 1796, and the Kant-Laplace

*Angular momentum is the product of mass, velocity, and size of a rotating body; left on its own, the body will conserve its angular momentum. If its size is shrinking, the body has to rotate faster to compensate for the decrease in size. The mass of gas and dust that surrounds a young star shrinks toward it while orbiting around it and attains a flat disk shape.

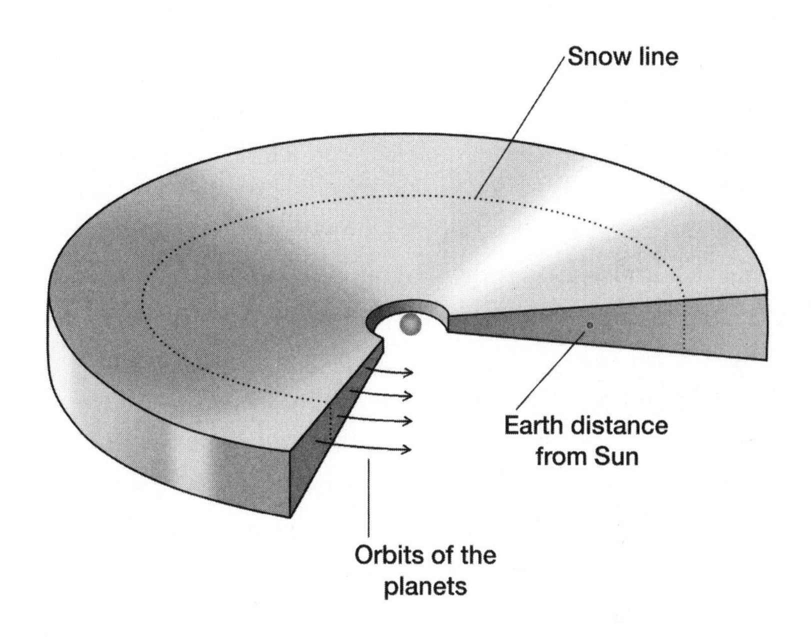


**FIGURE 1.1.** The newly formed star is surrounded by an orbiting disk of gas and dust, the material from which planets form. The disk is heated by the star and there is a curve at a distance where its temperature drops below freezing, known as the snow line. It is outside this line that snowflakes add to the dust in the formation of planets and help create gas giants like Jupiter.

model has survived 250 years of critiques, changes, and improvements while retaining its basic foundations.

There was something else that made me find Michel's discovery a bit hard to believe. According to the modern version of the Kant-Laplace model, there is a curve, roughly two to three times the distance of the Earth from the Sun, at which the temperature of the gaseous disk surrounding a star falls to just 170 Kelvin, or 150 degrees below zero Fahrenheit, at which point water and ammonia molecules in that rarefied atmosphere form ice grains and snowflakes.[2] These two light materials, as well as, ultimately, hydrogen, combine with dust particles and grow into giant gas planets orbiting the sun. Within the so-called snow line, dust particles, with no ice grains and snowflakes to aid their growth, combine to form small, dense planets (see Figure 1.1 on the preceding page). This is the beautifully simple explanation for the makeup of our solar system, gas giant planets orbiting the sun farther out and taking years to complete their journeys, and small, rocky planets orbiting closer in. So you can see why I was surprised by Michel's claim—there was no way a Jupiter-like gas giant planet could have ever formed inside the snow line. And orbiting 51 Pegasi, a star like our Sun, in just four hundred days—that just seemed impossible.

At the press conference the next morning, I found out I had been mistaken about the four hundred days.

It was four days!

Somehow my brain had locked onto the incredible figure and multiplied it by a factor of one hundred. Yet there was

Michel, with the evidence to back his claims, showing that the orbital period of the new planet was 4.2 days!

My deeply held preconceptions fell apart like ice grains and snowflakes meeting the Sun. It was a powerful—and humbling—lesson.

News of many more planets has followed the discovery of 51 Peg b.[3] Geoffrey Marcy and Paul Butler in California, already pursuing a similar project and technique, discovered several interesting planetary systems within months of Michel's announcement, allaying any lingering doubts that what Michel interpreted as planet 51 Peg b might have been an unusual property of its star. It was also easier to go back to an early find and accept it as a possible planet—the companion of star HD 114762, discovered in 1989 by my colleague and pioneer planet hunter, David Latham, and his collaborators.[4] It was also possible to see why the pioneers of the technique, led by Gordon Walker of the University of Victoria in Canada, had failed to discover a single extrasolar planet: they had done a systematic search from 1986 to 1995 but looked for planets with periods of ten years or longer, which limited the number of stars they could monitor. With some bad luck, the search ended empty-handed.[5]

Planets orbiting other stars, dubbed extrasolar planets or exoplanets, now number in the hundreds—about 600 at the time of this writing. All of them lie in our Milky Way Galaxy, relatively close to home, most within a circle of 500 light-years, although a handful are as far away as 5,000 light-years. More than sixty of these planets are similar to 51 Peg b

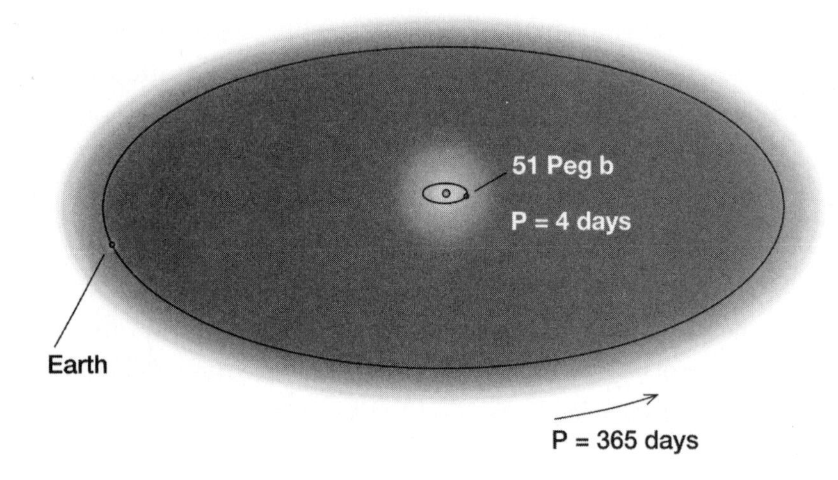


**FIGURE 1.2.** The orbit of the first "hot Jupiter" planet, 51 Peg b. The two orbits are shown at the same scale. The distance from the Earth to the Sun is 93 million miles, from 51 Peg b to 51 Peg, 5 million miles.

and are referred to as "hot Jupiters" (Figure 1.2). This number, which is fairly high, reflects the fact that the planets are easy to find, not that they are numerous. These planets, which at first seemed so anomalous (how could they have formed so close to the heat of their stars?), ended up having an explanation that didn't require throwing out the Kant-Laplace model. The hot Jupiters opened our eyes to the phenomenon of planet migration, the result of slow changes to a newly formed planet's orbit due to its interaction with the disk of gas and dust. As the orbiting planet raises density waves in the disk, its orbit can spiral inward or outward. In most cases, the shift is inward; the result is hot Jupiters.[6]

So while my experience in the beautiful old city of the Medicis took some time to sink in, when it did, I was deeply inspired to find answers to the questions that just days before I had taken for granted.

Thirteen years later Michel and I met again at the same conference. This time Michel described a bounty of small planets, perhaps like Earth, that he had discovered. I reported, based on computer calculations, on the strange worlds some of them might be. These smaller planets are more numerous and diverse than anyone had expected—searing hot planets with iron rain, atmospheres with 1,000 mile an hour winds, planetary systems with two suns, a planet that literally skims the surface of its star once every three months, and more.

Today we stand on the threshold to new worlds—planets that we could call home, planets that someone else might call home already. The search for them has spawned a new space race: the race to discover an Earth twin planet. The zeal and effort going into this race may seem odd and unjustified. Even for scientists there is no overwhelming benefit in discovering an Earth twin, because to study the properties of Earth-like planets they could rely on bigger ones, which are much easier to find. Yet everyone agrees that this is a historic moment. What gives rise to the extraordinary excitement of this race is the human yearning for meaning and belonging. It is the twenty-first-century version of the age-old question of "the Other," but on a grand scale.

The question of the Other is about how a conscious human being perceives his own identity: Who am I and how do I relate to others? It arises front and center during first encounters. Human history is full of first encounters: *Homo sapiens* encounters *Homo neanderthalensis* somewhere in today's Europe, Mayans encounter Spanish conquistadors in Central America, and so on.[7] The time of first encounters on our planet is now over. For better or worse, we humans—all of us—know about each other. The present generation of *Homo sapiens* has a global awareness, a sense of social connectedness, and an understanding of a common genetic makeup. The end of the twentieth century marked a real watershed in this sense.

The discovery of new worlds orbiting distant stars offers a fresh opportunity to contemplate a first encounter. As in the past, humans approach it with both insatiable curiosity and fear, with mixed, very strong emotions. As in the past—amazingly, despite all our modern technology and the visions of *Star Trek*—the new worlds we have just begun to uncover are enshrouded in mystery and full of surprises. And we will never stop exploring, as T. S. Eliot famously wrote: "We must never cease from exploration. And the end of all our exploring will be to arrive where we began and to know the place for the first time."

CHAPTER TWO

# THE WORLD OF PLANETS

In the mid-1990s the world of planets was a small one that comprised the nine planets of the Solar System; Pluto's "planethood" had not yet been challenged. Still, those planets represented a diversity of environments not imagined. The cameras aboard the flotilla of spaceships exploring our Solar System had shown us those exotic places and taught us the basics of comparative planetology. We couldn't be sure just how important planets, let alone life, were to the Universe at large. Today, for the first time, scientists can look at both planets and life as integral parts of the Universe and its history. In what follows we'll do just that.

The planets in our Solar System form two groups—the gas giants and the terrestrial planets. Jupiter, by name and by physical stature, is the ruler of them all. How the ancients could have sensed that is a mystery, since it took scientists more than two and a half millennia to measure Jupiter's mass and size and confirm its enormity for a fact.[1] Four hundred years ago in Padua, Galileo Galilei first used his unusual optical device, the telescope, to look at Jupiter. Galileo saw a planet, not a point-like star, orbited by four moons. He called them stars—Medicean stars, naming them for his Florentine patrons, the Medicis—the distinction between stars and planets not having been clarified yet. Now we call them the Galilean moons of Jupiter; their orbits help us discern the planet's gravitational pull and measure its mass.* That measurement was one of the triumphs of Isaac Newton's law of gravity in the generation of scientists that followed Galileo, and it inspired young thinkers like Immanuel Kant to figure out how the planets formed. It showed that Jupiter had the mass of more than 300 Earths and more than two times the mass of all the other planets taken together. Jupiter is a giant indeed, rivaled only by its distant second—the ringed planet Saturn.

Jupiter is a gas giant planet—we know that from its average, or mean, density. Its radius is ten times larger than

---

*Here I use the word "mass." The mass of an object measures the amount of stuff (atoms or matter) in that object. Mass and gravity are interrelated: the strength of the force of gravity depends on the mass—a more massive body exerts a stronger force (or pull). Colloquially, we often say weight or heavy, instead of mass or massive. This is imprecise and leads to confusing statements, so I'll insist on saying mass and massive.

Earth's, which makes Jupiter's volume a thousandfold larger. Given that Jupiter has only 300 times more mass, it must be made of stuff that is less dense than our rocky Earth. Indeed, Jupiter and Saturn are composed mostly of hydrogen and helium, the two most common and lightest elements in the Universe, very similar to the makeup of the Sun and the stars (Figure 2.1).

The largest planets in our planetary system resemble the Sun in another important way—they have no solid surface or geography. From the top of the atmosphere that we see going down, it is all clouds and more clouds, getting denser and hotter as we sink deeper. Most of Jupiter's interior is hydrogen and helium under pressures a million times higher than we are used to on Earth. One reason why the pressure inside is so much higher is that the larger the planet the stronger it pulls itself together by its own gravity—you and I would weigh 2.4 times more on Jupiter. If we were to venture deeper inside the planet, like diving in the ocean, the pressure would become higher as well. No wonder, then, that things can get a bit out of hand inside Jupiter. The hydrogen gas turns into a liquid known as metallic hydrogen. It conducts electricity, which is why we call it that; otherwise the substance has the least bit of resemblance to the copper wire in your bedside lamp. Studying the properties of this exotic material is a challenge in a lab—it was produced on Earth about ten years ago. Today we know it sufficiently well to describe—in computer calculations—more or less confidently the interiors of Jupiter and Saturn, and consequently hot Jupiters like 51 Peg b as well.

Both Jupiter and Saturn have a small core (small for them, but enormous by Earth standards) made of elements heavier than hydrogen, helium, and neon. A core is typical of a planet, left over from its birth and formative years. For comparison, stars are born without cores and live long without them. As they age, stars grow a core, as lighter elements are fused into heavier ones, which simply pile up inside the star. Surprisingly, Saturn's core, with a mass of about fifteen planet Earths, is bigger than Jupiter's, at three to ten Earth masses. Or at least we think so. Jupiter is so much bigger, with so much metallic hydrogen, that the content of its core is difficult to determine. If its core indeed turns out to be smaller than Saturn's, that could have happened by birth, or it could have eroded slowly and gotten mixed into the upper layers. More importantly, both Jupiter and Saturn have a similar fivefold excess of heavy elements compared to the Sun in their core and mixed in throughout, revealing in no uncertain terms their planetary ancestry.

Uranus and Neptune are a different story. While Jupiter and Saturn, like the Sun, are mostly hydrogen and helium, Uranus and Neptune have only 10 percent of their mass in hydrogen and helium. The rest contains lots more oxygen, carbon, and nitrogen, in the form of frozen water, ammonia, and carbon dioxide. Although ten to twenty times less massive than Jupiter and Saturn, they are giants compared to Earth, and so they are known as the ice giants. Pluto is compositionally quite similar to Uranus and Neptune, but much smaller.

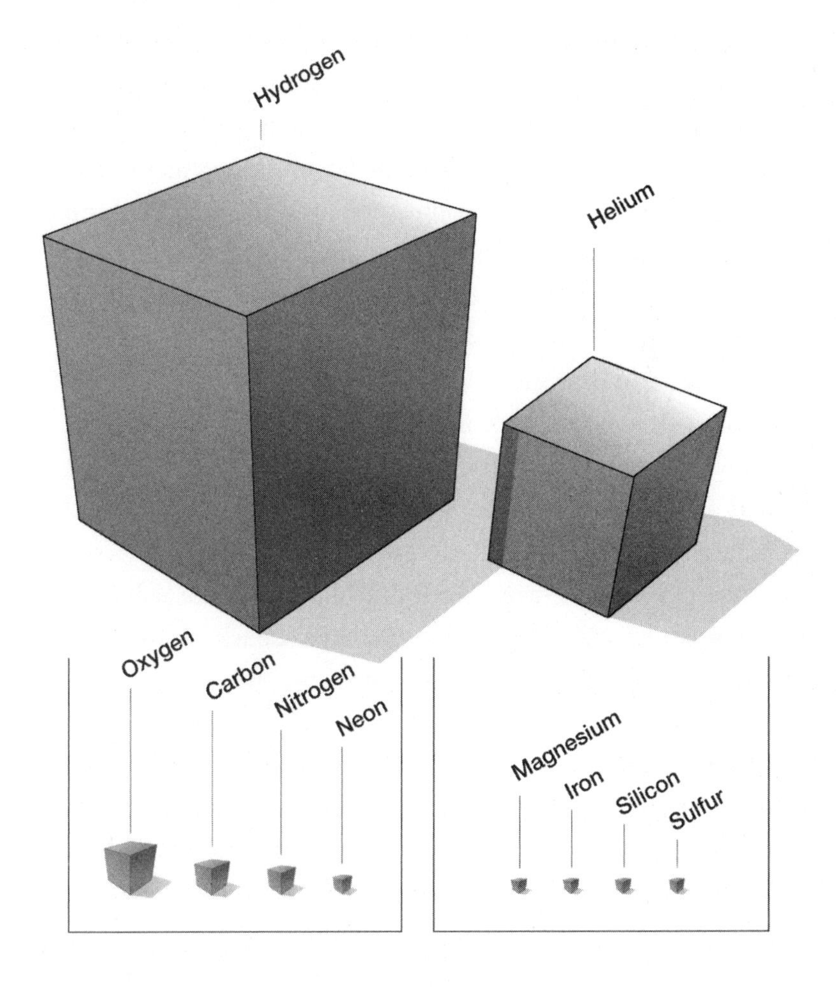


**FIGURE 2.1.** Proportions of the most abundant elements in the Universe today. This is the makeup of our Sun and most of our Milky Way Galaxy.

Much closer to the Sun is the province of the terrestrial planets, where Earth (a.k.a. Terra) rules in size and mass over Mercury, Venus, and Mars. Here hydrogen is almost gone—less than 0.1 percent by mass—and helium is virtually non-existent.[2] The terrestrial planets are mostly oxygen, iron, and silicon, although iron predominates on Mercury. Most of the iron in these planets resides in central cores. During the planets' formative periods, iron (and a few other metals, such as nickel, that could not be part of the rocks) precipitated in large droplets in the center of the planets. The opposite is true of water: some is bound in rocks, but the rest, rather than sinking, stays at the surface. If the temperature and atmospheric pressure are right, a terrestrial planet will have liquid oceans.

An obvious question, given how different these planetary groups seem, is whether they could have come from the same "stock." The modern Kant-Laplace model teaches that planets form from material left over from the making of the star, which consequently ought to have the same proportions of heavy and light elements. Images taken with the Hubble space telescope and the infrared Spitzer space telescope show that planet-forming disks are just 1 percent as massive as their stars, and that less than 2 percent of that mass is in all the elements heavier than hydrogen and helium. So why don't the small planets have any hydrogen or helium? Because of their mass. It takes the gravitational pull of a very large planetary seed to catch and keep those light gases. Smaller planets, such as Earth or Pluto, just can't hold them, and the intermediate-mass ice giants formed so far out in the disk that they could only grow slowly. By the time they were ready to catch the hydrogen and helium, it had all but dissipated.

Now we can project the knowledge we have gained about these planets onto the planets discovered around other stars. We have seen among them Jupiters and Saturns, with small or large cores, and we have seen Neptunes as well. And we have seen more diversity than we ever imagined. As we hone our techniques to discover and study smaller planets, we are in for more surprises.

In the hierarchy of structures and objects in the Universe, planets occupy a place at the bottom of a sequence that starts with clusters of galaxies and continues through galaxies and stars. All of these structures assemble and develop under the pull of gravity—their own weight keeps them together. All except the planets have similar compositions that are dominated by hydrogen and helium. Thus planets, in breaking with this uniformity, are more than just the products of gravity: they present the richness of form that the full table of elements—chemistry—can afford.

Imagine a planet that is larger and more massive than Earth but smaller than Uranus. Would a planet like this have deep water oceans—being a true water world—or would it be a dry planet with huge volcanoes billowing smoke high into a thin atmosphere? This is what we are about to explore. First, however, we have to find them.

CHAPTER THREE

# COMPLETING THE COPERNICAN REVOLUTION

In 1543 Nicolaus Copernicus set in motion events that transformed science and, through technology, human society. His insight—simplifying the architecture of the cosmos and placing the Sun, not Earth, in the center of the planetary system—was essential to the scientists of the next two generations (particularly Galileo and Newton) and the creation of modern physics. The Copernican revolution went directly to the heart of the question about humankind's place in the world. Many thinkers, most famously Dutch physicist Christiaan Huygens (1629–1695), jumped from the Copernican view of Earth as just another planet to the possibility of life on other planets.[1] In 1686 Bernard de Fontenelle popularized the possibility of extraterrestrial life in *Conversations on the Plurality of Worlds*, and it reached a culmination 300 years later in books and movies like *War of the Worlds* and *Star Trek*.

Ironically, these conjectured other planets did not materialize for 450 years. Even the nearest stars turned out to be very, very far away; discerning the tiny planets that may orbit them required four centuries of technological development. Now we are finally within reach of completing the Copernican revolution by discovering analogs of the Earth and the Solar System.

What makes extrasolar planets difficult to find is their distance and the fact that they are orbiting stars that are far bigger and brighter than they are. Typically a star is 1 billion to 10 billion times brighter than any orbiting planet, at least in visible light. This is a huge contrast ratio. To make things worse, since the observer is far away, star and planet appear very close to each other in the telescope. Taken separately, the high contrast ratio and the apparent closeness of star and planet are solvable. Together, they have been nearly intractable.

The telescopes that have been in operation during the past twenty years, including the Hubble space telescope, are capable of collecting light from objects fainter than 10 billion times the brightness of the nearby stars. This is done in the same way that a photographer takes a picture at dusk—by keeping the camera's shutter open longer. Taking a longer exposure allows more light to accumulate on the detector inside the camera, revealing very faint objects. The famous image of the Hubble Deep Field was obtained by taking a thirty-three-hour exposure in visible light, revealing thousands of distant galaxies.[2]

Many of the extrasolar planets known today could be detected by a very long exposure like that, except that the star makes a huge bright smudge in the middle of the image. The star will be "overexposed," as a photographer would say, and its light would be scattered all over the image. In fact, it could even damage the detector. Somewhere, lost in this scattered stellar light, is the faint light speck of the planet. That is why discovering a planet orbiting a normal star is such a big challenge.

Solutions have been proposed.[3] One method is to try observing the star and planet in other types of light. The star-to-planet ratio might be a billion to 1 or 10 billion to 1 in *visible* light. But as we know, light is a mixture of colors—waves of different length (or wavelength). These waves, when spread out according to wavelength (as done by a prism, for example) comprise a spectrum, as when water droplets turn sunlight into a rainbow. Consequently, applying a prism and looking for wavelengths in which the ratio between star and planet isn't so great might help.

This does work in some cases. For very hot planets, such as 51 Peg b, the star-to-planet contrast ratio improves a thousandfold (down to $10^7$) when observed in infrared light. Infrared is light of longer and longer wavelength, beyond what our eyes see as red light; our skin detects it as heat. A hot planet stands out better in infrared light next to its star because it "shines" with its own heat. A hot Jupiter can have a temperature of 1,500 to 2,000 K, which is much hotter than Earth (at 287 K) but is comparable to the Sun (at 5,800 K). Nevertheless, the $10^7$ contrast ratio is still daunting. Recently, infrared observations of known extrasolar planets have succeeded in special cases, but they still don't yield images, and the method is still not used for discovery.[4]

What other star-to-planet comparisons could we exploit? First, there is mass and then size; for both of these the star-to-planet ratios are much more favorable. For example, the Sun is 1,050 times more massive than Jupiter—so their star-to-planet mass ratio is $10^3$. That is much more manageable than $10^7$. With sizes, things get even better—the Sun is "just" ten times the size of Jupiter (and just 109 times the size of Earth)! This sounds good in theory, but how can we use it in practice?

Let's look at the star-to-planet mass ratio, since methods that exploit it have been the most successful and popular so far. The mass of an object determines its gravity—a more massive body exerts a stronger force (or pull). Thus the Sun makes Jupiter revolve around it in an eternal bind. But wait! Is Jupiter orbiting around the Sun like an anonymous slave, or are Sun and Jupiter waltzing their way through the Galaxy?

A waltz it is! To every force there is an equal and opposite reaction force, so the Sun and Jupiter balance each other around their "center of mass," which is a virtual point that is always on the line that joins them. They *both* orbit around the center of mass, just like a dancing couple. The Sun, being a thousandfold more massive, keeps their center of mass very close to itself, yet that virtual center is not inside the Sun. The Sun-Jupiter center of mass is about 7 percent of the solar radius above the surface of the Sun. To a careless observer from a distant star this might seem indistinguishable from Jupiter just revolving around the center of the Sun. But an astute and observant astronomer would see the waltz (or wobble, if you will) of the Sun as it orbits around the center of mass with Jupiter. The beauty of this trick is that the astronomer could observe the wobble of the Sun even if unable to see Jupiter in any other way! This is an indirect method of discovering a planet.

There are several practical ways to exploit the star-to-planet mass ratio in order to discover extrasolar planets. Three of them make use of detecting the wobble of the parent star and one exploits the mass ratio in a snapshot of sorts. A star's wobble can be detected directly by carefully observing the position of the star with respect to other stars over a period of time longer than the orbital period of the putative planet. This method—astrometry—allows us to isolate influences on a star's behavior caused by orbiting planets, and not by the Universe at larger scales. It has been tried for many years, at least since the early twentieth century, but turned out to be very demanding. Recently, the Jet Propulsion Laboratory at NASA developed the technology needed to achieve astrometry at the required precision from space, so the method might still deliver in the future.[5]

Another wobble method delivered the first new extrasolar planets. It relies on the influence of the Doppler effect on the spectrum of light coming from a star. The Doppler effect, as described in 1842 by Austrian physicist Christian Doppler, is something you are sure to have experienced. It happens anytime an object is moving and emitting waves at the same time. Sounds are waves in the air that we hear, so the sound of a passing car (or an ambulance siren) changes its pitch because of the Doppler effect: you hear a slightly higher sound as the ambulance approaches, and a slightly lower one after it passes by. Moreover, it is the relative motion that matters. So if the ambulance is stopped with its siren on and you drive by it, you'll experience exactly the same Doppler effect. The Doppler effect can be used in a practical way to measure the speed of objects, as, for example, when police radar catches you speeding on the highway, or when meteorologists measure shifting wind speeds in an attempt to detect tornadoes.

Light also consists of waves—electromagnetic waves—that are shorter than sound waves. So, light is subject to the same Doppler effect as sound, and we can measure the relative speeds of stars, galaxies, and more. This is of huge importance to astronomy, where we have no way to approach these distant objects and measure their motions.

To use the Doppler effect, scientists employ a telescope, a prism, and a camera to record the light spectrum of a star. Then they look for a marker in the spectrum. The marker will be shifted to longer wavelengths (toward the red light) if the star in question is moving away from you—the faster the star is moving, the more the marker is shifted. If the star is approaching you, the reverse is true. The majority of the visible light coming from a star is just like what we get from the hot filament of an incandescent lightbulb, and when dispersed into a spectrum it looks like a rainbow. But when a scientist looks at it with a prism, for example, she sees lots of markers. In the spectra of most stars these markers are narrow dark lines which indicate that light at specific wavelengths is missing. These lines—spectrum absorption lines—are caused by atoms and ions near the star's surface. When light passes through a gas made of atoms, for example, some of it gets absorbed by the atoms; we see this every day with clouds. However, our eyes are not equipped to see that the light is also absorbed in numerous specific wavelengths, or colors, of that light. These absorption lines and their corresponding wavelengths are due to the electrons that orbit atoms (and ions and molecules). Electrons have numerous, strictly defined states of energy—different orbits, if you will—and that is where light gets lost. Because these energy states are strictly defined, so are the absorption lines and their wavelengths. Therefore each atom of every known element—from hydrogen to the heaviest known metals—has an unmistakable fingerprint consisting of thousands of absorption lines all over the spectrum.

These spectrum absorption lines are the markers we use to measure the Doppler effect in stars. In doing so, we can discern very small changes in the speeds of stars, including the wobble that an orbiting planet would cause. What is more, the Doppler effect measured this way will allow us to measure the actual speed of the star on *its own* orbit around the center of mass. That orbital speed, together with the orbital period—which is the same for star and planet as they revolve around their center of mass—will allow us to measure their masses. We know how to estimate the mass of a star by measuring its spectrum; these absorption lines carry a lot of information. Therefore, we can determine the mass of the planet. There is one small hitch: we can't measure the angle at which the orbit is inclined toward us, as the Doppler effect gives us relative motion only. Consequently some uncertainty about the mass of the planet remains. To remove that uncertainty, we need to turn to other methods of planet discovery and study—for example, the transit method (more on that in the next chapter).

The third practical way to detect the wobble is by timing. If there is a strictly periodic signal we can measure and time with a precise clock, the wobble due to a planet will appear as a cyclic variation in the period of that signal. What could such a periodic signal be? Well, it could come from a pulsar—a neutron star spinning very quickly and emitting radio pulses—or from two stars orbiting very close to each other and eclipsing periodically (every few hours or days). That is how in 1992 the planets orbiting the pulsar PSR 1257+12 were discovered by the radio astronomers A. Wolszczan and D. Frail.[6] This was a remarkable discovery, but it did not gather the attention lavished on the 1995 discovery of 51 Peg b for two reasons. First, the pulsar planets were exotic—both as planets and as a planetary system.[7] Second, it turns out that pulsar planets are extremely rare. With only a couple of systems known, there is not much to study and little to help us understand their origin.

The timing technique can be used also when one planet is pushed and pulled around by a second planet. For example, if we have a system with a hot Jupiter and keep observing the hot Jupiter as it regularly orbits its star (e.g., by marking events like eclipses or transits; see the next chapter), we can catch variations in its regular orbit caused by an unseen second planet.[8] In some cases, the second planet is easier to catch by such timing variations, than by seeing the corresponding wobble of the star. This technique has become one of the big early science successes of the Kepler planet search mission. For example, five transiting planets in the Kepler-11 system could be confirmed and their masses derived by using transit timing variations alone.[9]

There is another method of planet discovery that also exploits the favorable star-to-planet mass ratio—gravitational lensing, an effect predicted by, and famous for its help in confirming, Einstein's general theory of relativity. The pioneering theoretical work of the late Princeton astrophysicist Bohdan Paczynski showed the practical uses of Einstein's prediction and led to the creation of several international projects to monitor stars for gravitational lensing.[10] For this, a source of light is needed, usually another star, behind the star being investigated. As we look at it, the light from the background star will be bent by the gravity of the intervening star. If the intervening star has an attendant planet, this will alter the lensing effect in a noticeable way.[11] In 2005 J. P. Beaulieu and his team discovered a planet with about five to six times the mass of Earth, which we call a super-Earth.[12] The planet bears the impossibly complex name OGLE-2005-BLG-390Lb, and it orbits a small star at a distance at least two to three times greater than the distance at which our planet orbits the Sun.[13]

That is probably as much as we will ever learn about planet OGLE-2005-BLG-390Lb, because the gravitational lensing method allows just a single glimpse, a sort of a snapshot. By its nature the observation cannot be repeated.[14] However, the value of the gravitational lensing method to extrasolar planets is in the statistics. In essence, the method is a general scanning approach that allows the discovery of super-Jupiters and super-Earths on an equal footing. So, even though very few planets have been discovered with this method so far, it was possible to notice a statistical trend that smaller planets are at least as common as giant planets, and probably are even more numerous.[15]

The first decade of extrasolar planets saw the maturation of several methods of discovery. Some of them complement each other and also help us study the planets we discover. This is crucial in our quest to find out if the Solar System, planet Earth, and—ultimately—Earth life, are unique, rare, or common in the Universe. Once we have achieved that, we will have completed the Copernican revolution. The last of these, which exploits the size ratio of planets and stars, and is known as the transiting method, seems to be the easiest. After all, the inequality between star and planet is the least daunting, with a factor 10 to 100 difference in size. And indeed, in our quest, transiting plays a special role. But, as often happens, there is a catch.

CHAPTER FOUR

# CHASING TRANSITS

It is about 5:30 in the morning and I am racing down an empty avenue by the Charles River in Cambridge, Massachusetts. Crossing the bridge into Harvard Square, I keep an anxious eye on the eastern horizon, where the rising sun is competing with clouds for a piece of the sky. My destination—the Harvard University Science Center, just north of Harvard Yard—is a bizarre scene. By the entrance, the Harvard marching band plays an obscure piece over and over again, as hundreds of people try to get to the roof. Everyone is here to see a sight not seen by humankind for 122 years. It is June 8, 2004, and a transit of Venus is under way.[1]

The transit—during which the little black disk of Venus passes in front of the big glowing disk of the Sun—is one of those spectacles in the sky that lets you "see" the inner Solar System as if it were a mobile. Transits of Venus are spectacular but rare: only every *other* generation can witness them. This generation is lucky: the transits of Venus now happen in pairs and the next one will occur on June 6, 2012.[2] Mark your calendar!

Today the transit of Venus is of little value to the research astronomer, but two transit cycles ago, back in the year 1769, the scientific rewards of witnessing such a transit were great, and international efforts to observe it were remarkable. According to William Sheehan and John Westfall, these efforts represented an eighteenth-century equivalent of the twentieth-century race to the Moon.[3] At stake was a unique opportunity to use the passage of Venus in front of the Sun to measure precisely the distance between Earth and the Sun, and thus distances across the Solar System. This was important for more than pure science, as the British Transit Committee duly noted in a memo to King George III in 1767; it was also crucial for navigation. That was enough to convince the king to support a mission. On May 25, 1768, Captain James Cook was appointed to do the job, and his famous trip to the South Pacific ensued.

A transit is in essence an eclipse of the Sun by either Mercury or Venus, although the event is not nearly as dramatic as when the Moon eclipses the Sun. Planetary transits do little to diminish the amount of sunlight we see—just a small fraction of a percent, too tiny for us to notice—but they do form a black dot against the Sun as they pass between us and the star. Historically, the term "transit" was reserved for Mercury and Venus, but these days it finds a much wider application in the hunt for planets around other stars. As with

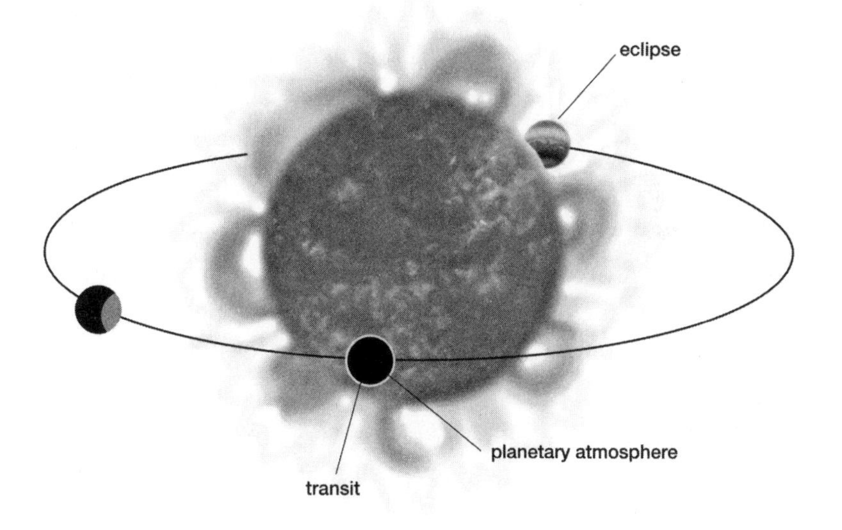


**FIGURE 4.1.** A schematic illustration of a transiting planetary system. The transiting planet is shown in three different positions on its orbit. When in transit, the dark night side of the planet is facing us, but we can glimpse its atmosphere in the stellar light that passes through it. When the planet is on the side (left side), it shows phases to us, just as our Moon does. Half an orbit after a transit, the planet will pass behind its star—known as occultation or eclipse.

Mercury or Venus, the planet will obscure only 1 percent or less of its surface (see Figure 4.1).

As it passes across the disk of its parent star, a planet will dim the star's light by a fraction equal to its projected area compared to the star's shining projected area. The area of a circle depends on its radius, $r$, squared. Therefore the star's light will dim by a fraction $(r_p/r_s)^2$, where $r_p$ and $r_s$ are the respective radii of the planet and star. For a planet the size of Jupiter and a star the size of the Sun, this dimming effect would be roughly 1 percent, which is easily detectable even with amateur equipment.[4] Earth, however, some 109 times smaller than the Sun, will dim just $1/(109)^2 = 0.008$ percent of the Sun's light when transiting. That is challenging to see, but not impossible.

The transiting method for planet discovery works because the dimming due to a planet transit can be measured by using the brightness of the star, known as photometry (measuring photons). Photometry is different from spectroscopy, the measurement of the color of light; for one thing, you need a camera to do photometry, and most often that camera is attached to a smaller telescope. The main difficulty is that a transit requires the planet's orbit to be almost exactly edge-on as we look at the star, which is very unlikely, as planetary systems in our Galaxy are inclined randomly in all possible directions (Figure 4.2).[5] Therefore, from our own vantage point, the probability that we'll see a transiting planet among the ones that are out there will be equal to the ratio of the stellar radius to the size of the planet's orbit.[6] This is generally less than 1 percent.

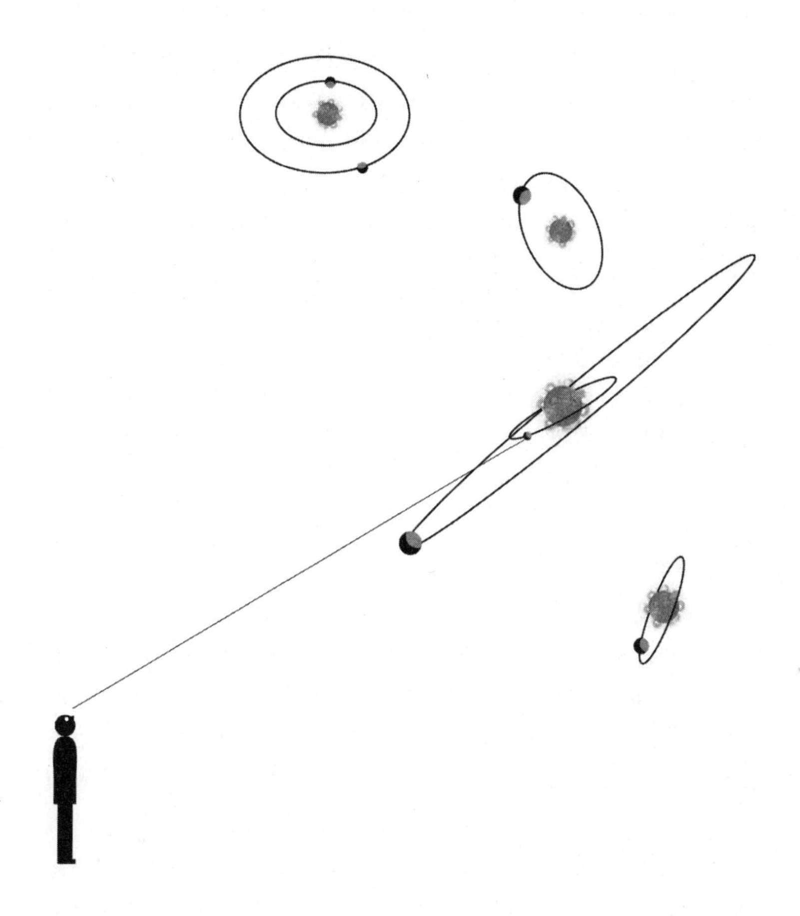

**FIGURE 4.2.** Planetary systems are inclined at random angles. When we observe them from Earth, few of them will be aligned so that we can see a planet transit. The closer the orbit of the planet to its star, the higher the probability that we can see a transit.

With such a low probability, tens of thousands of stars must be monitored patiently to detect periodic dimming due to planet transits in a handful. In this only the gravitational lensing method rivals the transiting method. Perfecting the transiting method and making it work in practice took a lot of work, but the effort was worthwhile because of the method's added benefits: when we can measure both transits and Doppler wobble, we can deduce the planet's radius and mass, and hence mean density. But that was just the bonus! The transiting method turned out to be the best path to discover really small planets, whether super-Earths or Earth analogs.

By 1999, more than twenty-five extrasolar planets had been discovered by the Doppler shift method, most of them hot Jupiters. Because hot Jupiters orbit so close to their stars, the probability that they will transit increases. Figure 4.2 shows how for the same planetary system an inner planet might transit, while another planet on a larger orbit would not. For such close-in planets there is a 5 to 10 percent chance that they will be seen in transit. In other words, for the more than twenty hot Jupiters discovered by the Doppler shift method, we should expect at least one of them to transit. The relatively high probability of finding a transiting planet makes the hunt for them a little more competitive. As planet-hunting teams discover new planets, they may keep their existence a secret until they check to see if the planet is transiting.

The lucky hot Jupiter turned out to be HD 209458b, an otherwise ordinary system of a planet orbiting a Sun-like star every 3.5 days, about 150 light-years from Earth. The planet was discovered by the Doppler shift method in the summer of 1999 by a collaboration of the Geneva Observatory and planet hunters from the Harvard-Smithsonian Center for Astrophysics. By September 1999 they handed their data to a Harvard graduate student, David Charbonneau, who was spending time in Boulder, Colorado, with the small photometric telescope setup built by Tim Brown of the National Center for Atmospheric Research. David and Tim detected a transit, and so did the team of Geoff Marcy and Paul Butler, who had been racing to do the same. They had handed their own Doppler shift data to Gregory Henry of Tennessee State University, who did the photometric measurements with an automated telescope in Arizona.[7]

This success was a watershed for two reasons: it confirmed beyond doubt the planetary nature of the extrasolar planets that had been found with the indirect Doppler shift method since 1995, and it boosted the effort to use transits as a method of discovery.

In the early 2000s the transiting method seemed to have a clear recipe for planet discovery: (1) do photometry of tens of thousands of stars simultaneously and (2) wait until you find a star that "blinks" in a regular fashion. If the star's light dims for a couple of hours by about 1 percent once every few days, then you have discovered a transiting hot Jupiter similar to HD 209458b. Two things seemed crucial: being able to measure a very large number of stars simultaneously and being able to do it with better than 1 percent precision. The former meant using either a small telescope that can see a lot of sky or a regular large telescope but measuring only faint stars. The latter meant mostly improving the software and the details of the photometric measurements.

Many astronomers rushed to discover the first planet with the transiting method. The expectations were high and the predictions were very optimistic.[8] Measuring the light dimming due to the transit was thought to be sufficient to confirm the planet. Consequently even teams with very limited resources could compete with Michel Mayor's team in Geneva and Geoff Marcy's team in California in discovering new extrasolar planets. The recipe was easy; reality turned out to be much more difficult. Three years passed after the discovery of HD 209458b with no new transiting planets.

As it turns out, the problem is that a few stars do "blink" regularly, but for the wrong reasons. For example, sometimes two stars in a close orbit would eclipse each other and a third star nearby would dilute the effect of that deep eclipse and cause it to appear shallow—say a 1–2 percent deep, as if due to a much smaller planet-size body. All three stars would appear as a single dot of light even in our best telescopes. Or a very small star would orbit a star slightly larger than our Sun and the eclipses would also be about 2 percent deep and difficult to distinguish from a planet transit. The list of different scenarios continues. The realization gradually dawned that these false positives were quite pervasive. David Latham, the pioneer planet hunter at the Harvard-Smithsonian Center for Astrophysics, was helping a couple of teams confirm possible transiting candidates with quick-look small telescope spectroscopy. Instead of planets, he kept uncovering false positives among the photometric transit candidates.

The problem came to a head in 2002. The OGLE team that we met earlier had equipped its telescope in Chile with a new large camera the year before. Before continuing with their primary experiment of detecting stellar gravitational lensing in our Galaxy, team members turned their telescope toward several patches of southern sky rich in stars and observed them nonstop every night for about four weeks, hoping to catch planetary transits. After a few months of dealing with the gigabytes of data that they gathered, the OGLE team found about sixty "blinking" stars. The "blinks" looked like planet transits, as the stars dimmed by just 1 to 3 percent; the trick was to make sure which, if any, were not false positives.

The photometric data obtained by the OGLE telescope alone was not sufficient to identify the kind of stars that showed the regular dimming. More information, such as stellar spectroscopy data or even distances from Earth, was not available because these stars were all faint and distant and previously unknown. The OGLE team had published its entire list of transiting candidates on the Internet (before publication in a journal) and invited the world community of astronomers and planet hunters to sort it out.[9] A true race began.

As soon as the first list of OGLE candidates appeared on the Internet, my younger colleague Kris Stanek walked into my office and challenged me to find planets orbiting the stars. He felt that the Harvard-Smithsonian Center for Astrophysics might have the resources and—most importantly—the expertise to pull this off. I agreed with him about the expertise; after all, the first planet showing transits, HD 209458b, had been our local success just two years earlier. But finding planets on the OGLE list needed a new approach. In fact, the entire transiting method of finding planets needed to be sorted out. The simple recipe of the 1990s—using photometry to look for blinking stars—had not produced any results.

The first hint of what needed to be done came from the OGLE list itself. A few of the transit candidates on it showed changes in their light between consecutive blinks. I had seen this many times before, but it had nothing to do with planets. When two stars, known as binary stars, orbit each other very closely—in orbits similar in size and period to those of the hot-Jupiter planets—the stars literally pull each other into pear-shaped forms. Their asymmetric forms mean that the shape of the surface we distant observers can see differs over the course of the orbit, and, as a result, the amount of light we see varies too. In addition, stars also illuminate each other, and that adds to the light variation. Now, if the two orbiting stars happen to be aligned just right, we also see the stars eclipse each other. There was only one problem—stars are large and their eclipses are very deep (so deep that the first such binary star was discovered by the unaided eye of John Goodricke in 1782).[10] The OGLE eclipses were ten times smaller; how could that be?

At this point my background in stellar physics came in handy. These OGLE candidates were certainly eclipsing stars, but with one of two possible differences from the typical binary pair. Either a third star was also in the picture or a very small star was in an orbit with a very big star. In the first case, the third star washes out the depth of the eclipse, making it appear shallow; in the second case the eclipse is shallow to begin with. One or two of the stars in the OGLE list even showed the telltale sign of a mismatched pair of stars, as there were barely visible eclipses due to the smaller star. The problem was figuring out whether the rest of the OGLE list was composed of similar eclipsing binaries as well.

The resolution to our primary problem—distinguishing false positives from real transiting planets—lay in fifty years of understanding of how stars work, known as the theory of stellar evolution. Stars of different masses have highly predictable temperatures and luminosities at any given age, and binary stars are of the same age by definition. Convinced that this basic stellar knowledge could be used to solve the problem, I began to lay out the steps needed to confirm that a blinking star was in fact being dimmed by a transiting planet.

I am a theorist—about stars—and although I love using telescopes and their instruments, I wasn't able to manage the OGLE challenge by myself. The problem needed a team. Guillermo Torres (a.k.a. Willie), an expert on binary stars and spectroscopic observation, was ready for the challenge, and we agreed that we should talk to one of our graduate students who knew how to do spectroscopy of faint stars, or their explosions, as it happens. That student was Saurabh Jha, who then was spying on very distant supernovae to understand dark energy. Saurabh was already excited about extrasolar planets; he had collaborated with our senior graduate student, David Charbonneau, in observing the transits of HD 209458b. In the meantime, Dave had already been building his own telescope for seeking out transiting planets, while working at Caltech.

Willie, Saurabh, and I got to work right away. Our new method involved multiple steps. After using photometry to identify the potential transits, we obtained a single low-resolution or medium-resolution spectrum of the star, in order to identify whether it was a massive star. If it was, we excluded it. If it was not, we obtained more spectra, to look for a large Doppler-shift wobble. A large wobble would indicate that the "transits" are due to a star, not a planet. Next, we used more sensitive instruments to look for a wobble due to a planet. That step required the largest telescopes on Earth, with the best spectroscopy possible. If we detected the small wobble, the next step was to analyze the spectra for distortions in the absorption spectral lines. If such distortions were not present, and no second set of spectral lines was visible either, we put all the accumulated observations together and compared them to the predictions of the range of stellar models with different possible configurations of foreground and background intervening stars. At the end of the day, if all this cohered consistently, the planet was confirmed, and its size and mass were precisely determined.[11] Willie and I also had to prepare the stellar and binary star models that we would need to analyze the systems and determine if they were stars or planets.

But at Harvard we had no access to a large telescope with a precise spectrograph for Doppler shift measurements of the very faint OGLE stars. Only the largest—the Keck telescope—would do, and the lion's share of observing time belonged to the partner institutions, the University of California and Caltech, that had built it. Fortunately, I had an ongoing collaboration on both Keck telescopes with colleagues at Caltech, the wizards of optical astronomy. At Caltech, Maciej Konacki—a young researcher working with my colleague Shri Kulkarni, and also a Pole (like the rest of the OGLE team)—was more than excited and ready to bring in the Keck at the last step of our transiting method. The Keck instrument—the old reliable HIRES spectrograph designed by Steve Vogt and used by Geoff Marcy to discover many extrasolar planets with the Doppler shift method—would be made available to us, and Maciej made sure we completed the crucial last step for planet discovery.

We had a busy summer, first doing the observations in Chile and Hawaii, where the Keck observatory sits atop Mauna Kea on the Big Island, and then analyzing the data as fast as possible in between. We were in a hurry not only because we knew that there was a race and the competition was fierce, but also because the core of our idea was to complete our steps in the right order, eliminating false positives along the way, and to make the best use of the precious time on Keck at the end. The results from the first few steps, done in Chile, were stunning: most of the OGLE transiting candidates were not transits and not planets, eliminating more than 90 percent of the "planets" on the OGLE list.

Fortunately we still had five candidates left for the observations at Keck. We were optimistic, but it was already clear that the initial high expectations, that more than half of the transiting candidates could be planets, were severely dashed; what's more, at least two more tests remained. It wasn't impossible that all the candidates would have to be struck from the list. In retrospect, we were very glad to see that our transiting method was working so far—for the first time I felt that we had a clear edge and were ahead of the competition, who were mired in a heap of false positives. If false positives had not been such a major problem, it was conceivable that any of the other competing teams would stumble on a planet from the list by chance and beat us by a month or less in announcing the first one!

The Keck observations went fine, and we seemed to have a clear winner among the four candidate stars we had observed: OGLE-TR-33. It was given this less than poetic name because none of the stars on the list had been observed before, so it was named for its place in the catalog of the team that first observed it. Star 33 had a clear wobble, with an amplitude that corresponded to a very large planet or, more likely, a brown dwarf, the name for a small failed star. Finding a transiting brown dwarf was almost as exciting back then as finding a large planet, so we rushed testing OGLE-TR-33, only to find that it failed the last test. We could not believe it—we had even started writing a paper to the journal *Nature*, while doing our test and models for a second time. Now we had to abandon it. In the meantime, another one of our top candidates had passed all its tests with flying colors. Initially we had neglected it because OGLE-TR-33 had a clear large wobble and had seemed an easier nut to crack. This star was OGLE-TR-56, and it looked like a Jupiter-mass planet.

As November was ending, we had finally discovered the nature of OGLE-TR-33: it was a system of three stars, two of which orbit each other very closely and eclipse each other, while a large star nearby, the big and brightest in the system, washes out the deep eclipse of the other two. The third star does not have a wobble of its own, but a large wobble of another star in the system (its spectral lines have a large Doppler shift) causes a small distortion in the spectral lines of the third star. Because the third star rotates fast and its spectral lines are broad, that small distortion was just enough to give us the impression of a small wobble, as if due to a planet that matches the shallow washed-out eclipse. OGLE-TR-33 was the ultimate tricky false positive![12]

Now we focused our full attention on OGLE-TR-56. It had passed all our tests, including the spectral lines distortion test that had uncovered OGLE-TR-33 as a false positive. We felt very confident that OGLE-TR-56b was a planet precisely because of our experience with OGLE-TR-33 and the other false positives our new transiting method had helped uncover. The method was working, and we quickly got a paper accepted for publication in *Nature*.[13] In the first week of January 2003 I flew to Seattle to present our discovery to the meeting of the American Astronomical Society. Just like

Captain James Cook 235 years earlier, we had crisscrossed the Pacific Ocean to catch a glimpse of a transit, and we had succeeded.

To top it off, OGLE-TR-56b was an exotic planet—a record holder in several ways: the shortest known orbital period (only twenty-nine hours), hence the hottest known planet (close to 2000 K), as well as the most distant extrasolar planet (at about 5,000 light-years from Earth). The exotic properties caught the attention of the media, while for us and the planet hunters the biggest excitement was that the transiting method for planet discovery was finally figured out. In short, the unexpected large fraction of false positives had been the major obstacle, and our set of tests and use of stellar models solved the problem. Within the next three years we and several other teams would use our approach successfully to confirm more than a dozen new transiting planets. The path to discovering a true Earth was now open. A new age of exploration was upon us.

The first age of exploration began in the fifteenth century. In 1484 one of the men whose efforts would define the era, Christopher Columbus of Genoa, was trying to convince King João II of Portugal to finance his expedition to cross the Atlantic. The king was reluctant, not because he thought Earth was flat, but because Columbus insisted that it was only 10,000 miles around the equator, and that the westward route to India and the Spice Islands would be short. Portuguese sailors (who, thanks to the support of Prince Henry the Navigator earlier in the century, had sailed up and down the Atlantic by the African coast) had estimated a much larger size for the planet, pole to pole, and had gotten a number much closer to the actual value of about 25,000 miles. (Back in the third century BC, Eratosthenes, a Greek mathematician, had already estimated the same size.) As a result, King João II did not fund Columbus, who then left for Spain. He had the wrong number, but luck was on his side and he stumbled upon the New World.[14] The Portuguese, in the meantime, went to India sailing around Africa.

The search for new Earths is no different. Since the late 1990s we have had the knowledge and the technology to do it. We have debated numbers and methods. Now we are sailing and waiting for the day when one of us will shout "Terra!"

Just like the Portuguese sailors, who started exploring nearby areas, planet hunters must do the same. The sailors would venture into the Atlantic and find islands like the Azores or farther down the coast of Africa, or perhaps even get an early glimpse of South America off the coast of modern Brazil. Our team also began small and cheap during the initial "gold rush" on discovering transiting planets. Most notable of our early stakes is the Hungarian-made Automated Telescope, known as the HAT project or network (HATNet, for short). HATNet is led and literally put together by a young colleague of mine, Gaspar Bakos. Gaspar came to the Harvard-Smithsonian Center for Astrophysics as a graduate student at the enthusiastic recommendation of one of my mentors (and Gaspar's mentor too), Bohdan Paczynski (1940–2006) of Princeton. The rationale for the recommendation was that Gaspar was the person to take advantage of a revolution in digital imaging (cheap CCDs, such as you can find in any digital camera) and precise image processing.[15] Together, the two technologies meant we could look at "all the sky, all the time." Or at least lots of the sky, very often. That, Bohdan believed, could bring a revolution in astronomy. As so often before, Bohdan turned out to be right.

There are many applications for the "all the sky, all the time" approach, but discovering planets by transits is an obviously good one. My colleague Robert Noyes and I thought so and convinced Gaspar as well. Thus HATNet was born on two continents and on a shoestring budget—with photography equipment for telescopes and amateur-grade CCDs, but with professionally designed and machined hardware (in Hungary, Gaspar's home country) and software.

HATNet comprises six small telescopes (not much different from the large cameras with zoom lenses used by professional photographers), four on Mount Hopkins in southern Arizona and two on Mauna Kea.[16] They are automated, following a cleverly written computer program that receives inputs (e.g., priorities for what should be observed) from the astronomers during the day; then they work all night like robots. HATNet shuts down in case it detects too many clouds or inclement weather, and Arizona communicates important updates to Hawaii. Remember that Hawaii experiences sunset and sunrise later than Arizona. Therefore, the two HATNet telescopes in Hawaii begin their work night later and take over fields that the Arizona telescopes can no longer see. By having "eyes" in Arizona and Hawaii, HATNet effectively extends its work night from twelve to fifteen hours, and can catch and see more transits.[17] HATNet was designed to discover transiting planets like Jupiter and Saturn at nearby stars. It has found thirty so far, with some of them (e.g., HAT-P-11b) the size of Neptune; what's more, HATNet and other projects like it have set off a transiting "gold rush."

Just as anyone could set off for California with a sluicing pan and some grub, the would-be astronomer of today can set herself up to find a new planet simply by maxing out a credit card. (I hope nobody does that literally.) It hasn't all been shoestring budgets, though, as the big guns, such as NASA and the European Space Agency, did not wait long to join the rush. Although a transit-hunting operation can be set up on a limited budget, the big agencies have a real advantage—they can get things lifted into space. Stars twinkle when seen from Earth because the air moves constantly, and not just in one direction; there are different currents at different altitudes. The air currents act as sheets with multitudes of lenses that shift the point-like images of stars, just like the sunlight playing on the bottom of a swimming pool. Much of this twinkling happens at high frequencies—about once every hundredth of a second. And, as it turns out, all this makes detecting a transit of an Earth in front of a Sun-like star practically impossible even with the largest telescopes. A telescope in space has its own challenges (although as the Hubble space telescope taught us, they are not insurmountable), but to find small planets, it's the only way to go.

The Europeans already had a mission in the pipeline that could easily accommodate the transit hunting. Hunting the *co*nvection and *rot*ation of stars (known as COROT, and referring to the famous pointillist impressionist painter), the telescope was to obtain thousands of images of stars, resembling pointillist paintings, in order to study their subtle light variations and help understand basic things about stars, like their rotation and internal convection. Of course, COROT could do an excellent job detecting planet transits in the process, with its name now standing for convection, rotation, and transits (CoRoT).

In the meantime, NASA had missed an early opportunity to fund a space telescope dedicated to discovering transiting planets, one with a stated goal to discover planets as small as Earth and determine how common planets like ours actually are. William Borucki of the NASA Ames Research Center in California had been trying to convince the agency that his experiment would succeed. Even a null result, meaning that no transiting Earths were discovered, would be meaningful, implying that planets like ours are very rare. NASA panels had turned him down before, but in 1999 Bill Borucki was assembling a crew to propose again; he asked me and a dozen more colleagues to join. The successful discovery of more and more exoplanets with the Doppler shift method was a powerful new motivation.

Even though I hadn't done much work on the problem at the time, I had first thought of discovering such planets in 1999, when a group of us, mostly at the Harvard-Smithsonian Center for Astrophysics, and mostly observers and engineers, came together to propose to NASA an innovative space telescope design for planet detection—with a square mirror, as opposed to a round one. My colleagues Costas Papaliolios and

Peter Nisenson had invented this unusual design in order to minimize stellar glare and allow glimpses of planets huddled close to their stars. With a team of about twenty and led by our experienced space mission scientist Gary Melnick, we prepared a detailed scientific and engineering proposal.

My job on the team was to work out what kind of planets our telescope might be able to discover. It seemed then that super-Earths were in reach. (I liked to call them super-Earths and super-Venuses for short, as it had been common in astronomy to use the adjective "super" for newly discovered or hypothesized objects that are larger in size or energy than known ones. For example, stars that are larger than giant stars are called supergiants, explosions that are stronger than novae are called supernovae, and so on.) The shorthand stuck, as you've probably already surmised.[18]

Finally, in December 2001, the Kepler mission, as it was now known, was approved. Seven years later, in March 2009, Kepler was launched from Cape Canaveral in Florida and two months later beamed down to Earth images and exquisite measurements from stars and planets hundreds of light-years away. Kepler is NASA's first mission capable of finding Earth-size and smaller planets within the habitable zone of stars similar to the Sun (a topic we will return to later). It is a fairly modest telescope, about half the size of Hubble, but with a lens that allows a wide field of view, captured by the largest ever camera built for a NASA science mission—a 95 megapixel monster that can image millions of stars in a single shot.

The NASA Kepler mission will use the transiting method to discover planets like Earth—of the same size and in similar orbits. The goal of the mission is to do so in a systematic and comprehensive manner, so we can find out what fraction of stars have Earth-like planets. The mission is designed to avoid most of the pitfalls of the standard transiting method by investing in several years of preparation—a comprehensive and carefully vetted input catalog of 200,000 stars that the Kepler telescope is going to search for planets. After checking most of these over the initial few months, Kepler is to settle on about 100,000 to 120,000 stars for the life of the mission, about three to four years. All the advance work should mean that we will get very few false positives among the planet candidates from the Kepler photometry; at the Kepler team we have a plan how to weed out the remaining false positives, using the same methods we applied to the OGLE catalog. As I write this the approach is clearly paying off; our preliminary evaluation is that the fraction of false positives is very small.

Ultimately, the transiting method is more than just a successful technique to discover planets, including Earth-like ones, as valuable as this may be. Transiting planets are the best ones for us to study in the short term. Because we know their masses and radiuses precisely, we can infer their bulk composition from the mean density. In addition, observing transits enables us to remotely analyze a planet's atmosphere. During a transit, the light of the star is colored by the presence of the planet's atmosphere; we can compare the spectra then to the spectra observed when the planet is not transiting, and use spectrographic techniques to identify the chemicals—such as water, methane, or carbon dioxide—present in a planet's atmosphere.

History repeats itself. In terms of their significance, there are uncanny parallels between the first observed transits of Mercury and Venus in the seventeenth century and the first transits of extrasolar planets observed today. Today, as in the seventeenth century, the first transit observation convinced any remaining skeptics of the reality of a big new concept. Today, as in the seventeenth century, the first transit observation brought in an unexpected result as well. Today, as in the seventeenth century, the first transit observation opened the door for an important future use of transits. Perhaps the only difference is that back then it took a century to take advantage of that important utility of transits; for us it is already happening.

It is no coincidence that Johannes Kepler, the astronomer, mathematician, astrologer, and mystic, should be the namesake of our mission to find new Earths. He discovered the laws that govern planetary motion—we still use the laws as he wrote them to calculate the orbits of the planets we discover with NASA's Kepler telescope. But that is not all. A less known fact is that Johannes Kepler was the first to predict and calculate accurately the transits of Mercury and Venus.

Johannes Kepler never saw a transit. He died exactly one year before the first transit he had predicted, that of Mercury in 1631. But at least one astronomer, Pierre Gassendi, heeded Kepler's call and observed the transit from Paris on November 7, 1631.

November weather is often cloudy in Paris; Gassendi was beset by rain and clouds the days before. He had planned to observe the sun before and after the predicted time of the transit, since Kepler had noted the large uncertainty in his calculations. On the morning of November 7 the clouds cleared briefly and Gassendi saw the tiny black dot of Mercury on the solar disk. He was projecting the sun on a white screen, as we would do nowadays at home or for public viewing. Gassendi knew how to distinguish Mercury's dark image from sunspots.[19] Just twenty years earlier Galileo Galilei had used his new telescope to discover sunspots and describe how they move and change; Gassendi was an avid follower of Galileo and his experimental methods, and knew it when he saw the tiny dot move with respect to the spots in the direction of Mercury's orbit.[20]

The observation of Mercury's transit in 1631 was important for several reasons. Though most scholars had adopted the Copernican revolution, the rest of the world had not. After all, it was just thirty years since Giordano Bruno's death at the stake in Rome, and much less since Galileo's own troubles with the Inquisition. The transit was one more success for the heliocentric system, similar to Galileo's observations of the phases of Venus in 1609. Kepler's prediction of the time of transit was precise to within five hours, which was amazing for his time, and was a success for the Copernican system and its predictive power.

However, what most excited both Kepler and Gassendi about the transits of Mercury and Venus was the opportunity to measure directly the sizes of these planets or of any planet, for that matter. Kepler thought he had "discovered" another law, namely, that the volumes of the planets are proportional to their distances from the Sun. Unlike his laws of planetary orbital motion, this one was based not on evidence but on harmonic proportion as part of Kepler's worldview, his "harmony of the spheres."[21]

Gassendi was aware of this and was surprised to see a very tiny Mercury. He was well prepared however, and made precise measurements of its size, which he dutifully reported to a skeptical audience. Kepler's harmonic law of planetary volumes made Mercury and Venus too big and Jupiter and Saturn too small, but the scholars of that time had a tough time accepting that Kepler was wrong. Eight years later, when Venus was first observed in transit, the same surprise was in order. It appeared smaller than expected, and scholars finally understood that planet size and orbit are not strictly related.[22]

When the hot Jupiter planet HD 209458b was observed to transit its star in 1999, some skeptics argued that the new class of wobbling stars might be a phenomenon other than planets. The transit of HD 209458b removed once and for all any such lingering doubts. But the observation (as with Mercury in 1631) brought a surprise as well—the size of the planet was larger than expected, namely, Jupiter's size.

Ten years later the inflated radius of HD 209458b and a good dozen other such planets remains a mystery. The expectation, as we've seen, is that planets with the weight of Jupiter consist almost entirely of hydrogen and helium. With hydrogen and helium being the lightest gases in our Universe, there is a maximum size a planet could have after compressing the gases under its own weight. The presence of anything heavier, such as oxygen or metals, would cause the planet to shrink.

Jupiter and Saturn obey that theory well, so it is very surprising to find several extrasolar planet analogs to Jupiter that are 30 to 50 percent larger in size. Even in the unlikely event that they are made of pure hydrogen, their size can only be explained by a strong, persistent source of internal heat. Hot Jupiters are all very hot, but so far scientists have failed to identify a way to account for such a heat source.[23] Once we find transiting Jupiters with the Kepler telescope in a range of orbits away from their stars and measure their radiuses, we may glean a clue as to what makes puffed-up hot Jupiters so . . . puffed up.

Johannes Kepler made one more mistake regarding the transits of Mercury and Venus: he predicted only one of the pair of Venus transits in the seventeenth century. Kepler predicted the transit of Venus on December 6, 1631, hot in the footsteps of the Mercury transit the month before. Our friend in Paris, Pierre Gassendi, observed the Sun for three days in a row, but in vain. It was not his fault: the transit was visible only in the Western Hemisphere, as Kepler had correctly suspected. What Kepler failed to predict, however, was that just eight years later—on December 4, 1639—Venus would transit again and this time would be visible from Europe. A young Englishman, Jeremiah Horrocks, uncovered Kepler's mistake and went on to be the first to see Venus in transit.

How could the sixteen-year-old Horrocks do a better planetary calculation than Kepler, the father of orbital laws? The answer lies in what was to become the overarching importance of Venus transits a century later—measuring the "astronomical unit" (the Sun-Earth distance). In Kepler's time that distance was not known; there were estimates, and Kepler adopted a very short one. Thus for an observer on Earth, the angle subtended by Venus on December 4, 1639, was too large and Kepler predicted that Venus would skim the Sun, not cross in front of it. Horrocks redid the calculation with newly published tables that had adopted a longer Sun-Earth distance. The transit was a go!

This little anecdote reveals an important point—Kepler's laws are correct and precise, but they give us proportions of how one planet's orbit measures with respect to another planet's orbit. In order to get an actual (absolute) distance, say in miles, one needs to know at least one orbit in miles. Therefore, if we know the astronomical unit, we can measure the distances to all the planets in the Solar System. Today, knowing the astronomical unit precisely is still important because we need it to measure the distances to other stars and the planets around them.

About twenty-five years after Horrocks observed the transit of Venus, James Gregory in Scotland proposed transits as a method to measure accurately the astronomical unit.[24] In the meantime, precise navigation was becoming an increasingly important strategic issue for the imperial powers of the day. The astronomical tables, whose precision depended on a reliable astronomical unit, were central to good navigation. The next pair of Venus transits in 1761 and 1769 was no longer left in the hands of an occasional astronomer in a Paris apartment or a house in the English countryside. Chasing Venus's shadow was now a matter of international competition.

In order to appreciate the significance of the transits of Venus in the eighteenth century, consider some of the people who were involved and the global reach of their well funded expeditions. Charles Mason and Jeremiah Dixon, who later went on to draw the famous Mason-Dixon Line between Pennsylvania and Maryland, led one of the British expeditions to measure the first transit. Captain James Cook led the British expedition to measure the second transit and later went on to explore much of the Pacific.

The transits of Mercury and Venus have all but lost their practical importance in the twenty-first century. However, our ships are still leaving port to chase transits—up in space and for alien planets. Johannes Kepler would be glad to see his name on one of them. Captain Cook would marvel at them and admonish us to explore new worlds.

CHAPTER FIVE

# SUPER-EARTH

## *A New Type of Planet*

We love our planet Earth. We should—it is our home, and there's no place like home. There can't ever be a better place than Earth. Plenty of serious science literature supports that view in an emotionally detached manner. It is often called the "Goldilocks hypothesis": the Earth is just the right size (not too big, not too small) and just the right temperature (not too hot, not too cold) for life to emerge here. Life is a rare thing. Perched on our little planet, we can't see any other out there, or at least not yet—so a certain dose of Earth-centrism seems justified.[1] Or is it?

Life is extremely resilient once it takes hold, but it requires rich chemistry, large energy sources, and stability, right from the beginning. The comparative planetology of our Solar

System makes it seem like those initial conditions are hard to come by. Earth seems perfect, whereas the rest have obvious defects. Mars is on the smallish side, lacks a substantial atmosphere and water, and is very cold (although we still hope to find life there). Jupiter is too big; its crushing pressures and element-poor environment make interesting chemistry impossible. The trouble with such a comparative analysis, however, is that it leaves out a crucial class of planets that, purely by happenstance, doesn't occur in our Solar System.

These are the super-Earths, which we'll examine in the next two chapters.[2] A super-Earth is a planet that is more massive and larger than Earth, although still made of rocks—perhaps with continents and oceans—and an atmosphere. There is no such planet in our Solar System, but we know that they must be common in other planetary systems.[3] Moreover, theory predicts that they might have all the nice attributes of Earth, and, in fact, provide a more stable environment on their surface. True super-Earths!

A super-Earth is a planet defined by its mass—between 1 and 10 $M_E$ (where $M_E$ stands for Earth mass). The Earth's mass is $6 \times 10^{24}$ kg, or a 6 with 24 zeros. That's pretty big, but the Sun's mass is $2 \times 10^{30}$ kg—a million times larger—lest we lose perspective.

I limit super-Earths to about 10 $M_E$ for a couple of reasons. To start with, we have not seen any super-Earths above that limit yet. We know the mean densities of about a dozen small exoplanets above that limit, and they are all made of gas and ice mostly, like Neptune. Second, when planets form in a proto-planetary disk (as in Figure 1.1), 10 $M_E$ is roughly the critical mass at which hydrogen gas can be swept in and retained by the growing planet, turning it into a Neptune-like giant. Of course, right now this range is somewhat approximate, and the upper limit to the mass of super-Earths could be anywhere between about 10 and 20 $M_E$.[4]

The first super-Earth was found by the serendipitous work of Eugenio Rivera, Jack Lissauer, and the California-Carnegie team in 2005. It orbits the small star Gliese 876 and is about seven times more massive than Earth.[5] It is also very hot because it orbits very close to its star—only seven stellar radii, or approximately 1.7 million kilometers, away! (For comparison, this is only 3 percent of the distance from our Sun to Mercury.) Another, much colder super-Earth that is about 5 $M_E$ was discovered by J. P. Beaulieu and his team at the Paris Institute of Astrophysics, unfortunately using the gravitational-lensing technique described in Chapter 3. In early 2007 Michel Mayor's team in Geneva spotted at least three planets orbiting the star Gliese 581; two of them are super-Earths with minimum possible mass of 5 and 8 $M_E$ each, and orbital distances that are 7 percent and 25 percent of the distance from Earth to the Sun, respectively.[6] A year later the same team reported several more super-Earths, some orbiting stars as big and hot as our Sun. The first transiting super-Earth was discovered by the CoRoT space mission in 2009: CoRoT-7b, a very hot small planet, probably similar to Gliese 876.

With so little opportunity for direct observation of these planets, little besides a planet's mass and orbit can be known empirically. So scientists turn to theory to answer their questions about what these planets are like. "Theory" is what practicing scientists in astrophysics and planetary science call the research done to explain results from experiments and observations, as well as research predicting phenomena and objects not yet seen. (It is shorthand, and it should not be confused with what is known as scientific theory, which includes things like Newton's laws of mechanics, which are so well established that we treat them as natural laws.) Theorists are scientists who do not conduct experiments and do not make observations. In the past they worked with pencil and paper, but these days they stare at a computer screen. They model a process or calculate the properties of an object by solving mathematical equations that describe them. In the history of science, Albert Einstein and Stephen Hawking are theorists; Ernest Rutherford was an experimentalist; Edwin Hubble was an observer. Sometimes the distinctions are blurred.

Although observations are limited, scientists can study super-Earths today by building theoretical models with a computer, the idea being that they will then be tested by observation. By comparing predictions and observations, we can refine our models even as we learn more about these unusual planets. This is no different from how science studied Earth's interior and the other planets in the Solar System less than 100 years ago. In fact, what we know today about Jupiter-like exoplanets is about equivalent to our knowledge of Jupiter in the 1950s.

Theorists have applied this sort of effort to super-Earths only since 2004. That they were not studied theoretically before 2004 is a matter of neglect. Theorists usually work on phenomena and objects that are known to exist. Why waste time on something that is not immediately available?

I am a theorist, so I have to accept some of the responsibility. When Harvard planetary scientist Richard O'Connell and I first talked about computing a model for a planet like Earth but twice as massive, it did not seem to be a difficult project. We were just excited to see what would come out. It was no disappointment—the models of these big rocky planets were very interesting indeed—and by the end of the year Diana Valencia, the graduate student Rick had recruited to work on this project, was ready to give a short report at the American Geophysical Union meeting. I was still incredulous that no one had done such models before, so I asked Diana to find out as much as she could by talking to colleagues after her presentation. She came back very encouraged. People were very interested, and no one had computed such models.[7] We were treading new territory, literally!

My continuing motivation to find super-Earths had been boosted by experimental success with the method I had worked on for confirming the presence of planets as they transited their stars. The number of confirmed exoplanets continued to grow. And even more exciting was the advent of Kepler.

As a member of the Kepler team and preparing for the mission, I was acutely aware that we were going to discover many—hundreds of—planets smaller than Neptune but bigger than Earth, yet we knew nothing about how their masses and radii should relate to each other. (This was the problem that had taken me to Rick O'Connell, a colleague but in the Harvard Earth and Planetary Sciences Department, in the first place.)

There are two criteria for calling something a super-Earth, which Diana, Rick, and I established: (1) it is between 1 and 10 $M_E$; (2) it is composed mostly of solids (such as rock and ice).[8] This may not seem like a great recipe for variety, and it is true that some of the planets we modeled seemed quite familiar. Far more often, however, we found planets that were exotic and novel.

Let's start with the familiar and take a journey to the center of the Earth. The rocky super-Earth cutout in Figure 5.1a is a good illustration to the way ahead.

The outermost layer is the solid crust. What it's made of depends in part on where we stand. On our own planet, if we begin our journey in California, the crust will be a layer of rocks rich in silica, like granite, that goes about 20 kilometers deep. If we begin our journey on the Pacific Ocean floor near the Hawaiian Islands, the crust will be a layer of basalt rocks denser than silica, such as olivine, that goes only about 5 kilometers deep. On average the Earth's crust is 30 kilometers thick, thinner than that, as we have seen, under oceans, and up to 60 kilometers thick under continental mountain ranges. Our journey to the center is 6,400 kilometers long, so the crust of 30 kilometers is a very brief introduction.

As we go deeper the temperature rises steadily, as miners know all too well, as does the pressure due to all the rocks above us. Consequently, below the crust is a region of partial melting (the lava of many volcanoes originates there) that quickly becomes the mantle, a thick layer of hot rock, often described as molten. This is actually a misnomer. Yes, on Earth's surface rock at 2,000 degrees flows like liquid from volcanoes, but under the enormous pressure deep in the mantle that same rock is more like cold honey: malleable and extremely slow to flow.[9] The mantle is also in a state that resembles boiling in slow motion (called convection)—the mantle is so viscous that bubbles in it take millions of years to float to the top.[10] However, this is short compared to the life of the Earth, so on a geological (or planetary) scale of time there is a lot of churning and mixing going on. The temperature at the bottom of Earth's mantle is about 3,700 degrees; that heat is what drives the flow toward the surface. Meanwhile, colder and denser rock near the surface sinks and flows down, dissolving completely in the process. We can see some of the effects of all this churning on the surface, as this convection pushes around the fractured pieces of the crust—known as the tectonic plates—slowly rearranging the continents.

The mantle is Earth's largest layer, some 2,860 kilometers thick, and takes up 84 percent of the planet's volume. The mantle is much denser than the continental rocks, and consists mostly of a mineral called perovskite: a dark dense rock rich in iron and magnesium that is more than 50 percent denser than granite.[11]

Below the mantle, some 2,900 kilometers below the surface, we encounter the core, consisting of pure iron and iron

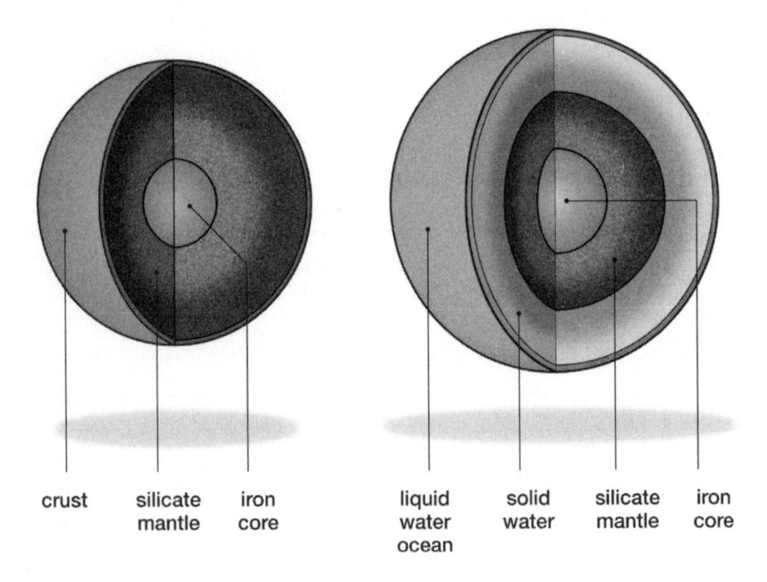


**FIGURE 5.1.** The interior of a super-Earth. The left image (a) shows the interior of a rocky super-Earth that resembles the interior of the Earth; on the right, the image (b) shows an ocean or water super-Earth with most of its water in a solid form.

alloys. The temperature here is very high—5,000 to 7,000 Kelvin—comparable to the temperature on the surface of the Sun. The pressure is very high too. The core is equal in thickness to the mantle, but, being in the center, it is only about 15 percent of the planet's volume. Earth's core consists of two parts: an outer liquid iron core and a solid one below it. Of course, the outer core is liquid in the same sense as the mantle is.

We have learned all this from studying earthquakes. The best view of the interior is afforded by analyzing the paths earthquake waves take as they travel through the Earth. Seismographs all over the planet record where and how fast the different waves produced by a single earthquake arrive after bouncing inside the Earth. (This is not too different from the way ultrasound images the inside of the human body from outside.) The full picture is put together by adding to this our knowledge of Earth's magnetic field, heat flow, gravity studies from spacecraft, and laboratory experiments on rocks under high pressure.

I alluded to how some of these structures form in Chapter 2, but the full picture is a bit more complicated than I've explained thus far. The preplanet structure—the "seed" of a planet—consists of solids (mostly silicates) and volatiles (such as water and ammonia), with trace amounts of hydrogen and noble gases. Due to the energy of the accretion process and the constant collisions with large solid bodies, this seed is thoroughly molten. (Some of Earth's internal heat is a relic of this process.) In this state the structure differentiates.

Iron and siderophile elements (high-density transition metals that like to bond with iron) precipitate from the silicate mix and sink under their own weight to form the core in the center. The remaining silicate minerals will remain in a mantle with the less dense ones closer to the top. Volatiles that are left over after hydrating the mantle minerals will rise to the surface and atmosphere.

This process, called planetary differentiation, is quick in geological terms and works for planetary bodies as small as big asteroids just a few miles across. Iron meteorites—pieces of pure iron alloy that orbit around the Sun until one day they fall to Earth for us to find them—originate in the differentiated iron cores of asteroids that were later smashed up by collisions with other asteroids. So, although we have no samples of our Earth's iron core, and no good prospects to get them anytime soon, iron meteorites are excellent proxies. Differentiation is an orderly and predictable process thanks to our knowledge of chemistry and mineral properties under pressure.

Some super-Earths, the rocky ones, develop quite similarly, although the pressure in the mantle is almost tenfold higher and different varieties of minerals form. Other super-Earths, the oceanic ones, are totally exotic beasts, with oceans that are 100 kilometers deep overlying a dense hot solid water, called ice VII.

It might seem ridiculous to refer to this water as ice, given that it is at a searing temperature of 1,000 K, but under such high pressures, it forms. Water—$H_2O$—has a familiar structure and formula, but our familiarity with it can make us overlook the fact that it is actually very complex. One key feature is that the oxygen atom in its molecule does not share electrons equally with its two hydrogen atoms; the result is that the molecule ends up with an asymmetric distribution of electrical charge. Imagine the tiny water molecules like small magnets, except with three poles (a negative O and two positive H's). Water molecules interact with each other because the positive charge near a hydrogen atom of one molecule bonds weakly with the negative charge near the oxygen atom of another molecule. Many such weak bonds together can form a strong structure if the temperature becomes low enough to allow it. Thus common water ice is formed, dubbed ice Ih or hexagonal ice. In common ice the weak bonds between the molecules cause the molecules to form rings (mostly hexagons) that leave lots of empty space in between. The empty space gives it a lower density than liquid water, and so—as you know from a glass of ice water—it floats.

Under high pressure the density of water increases as the molecules are forced closer together; the bonds are bent to form tighter rings, which also interpenetrate. That makes the water solid, almost irrespective of the temperature, and much denser than the liquid phase.[12] The high-pressure ices that exist at high temperatures are known as ice VII, X, and XI; these are the ice phases we expect to find inside oceanic super-Earths.[13] These ices are still less dense than rocks, however, so an oceanic super-Earth will be less dense than a rocky one of the same mass.

Ocean planets might be very common in the Universe because water is very common in the low-temperature environments where planets form and evolve.[14] This might be especially true for super-Earths, which can retain volatiles more easily thanks to their larger mass and surface gravity.[15] In order for a planet to become an ocean planet, it should form with or obtain at least 10 percent of its mass in water. Ammonia could be mixed in, but water is by far the dominant volatile chemical we see among the materials in protoplanetary disks. For comparison, Earth's oceans are just about 0.02 percent of its mass. However, a much greater amount of water could be incorporated into Earth's mantle.[16] For that reason I assume a much larger fraction of water (greater than 10 percent) to produce a separate uninterrupted layer of water surrounding a planet (Figure 5.1b).

An ocean planet, regardless of its surface temperature, should have the same layers inside: an iron core surrounded by a silicate-rich mantle that transitions into the hot water ice. The latter will become liquid water near the surface (the last 100 kilometers or so). The surface of the liquid water ocean will be covered with ice Ih, if the planet is far from its star and cold, like Jupiter's moon Europa. If the planet is close to its star and hot at the surface, the liquid ocean will transition into a thick hot steam atmosphere. If the planet has moderate temperatures such as we have on Earth, the water ocean will resemble Earth's, but there will be no continents or basalt tectonic plates under it. The interior of the ocean planet will remain under the control of the planet's internal reservoir of heat. The transition between silicate mantle and hot water ice happens with a small change in density but no change in temperature, and the two materials have similarly high viscosity. Like the silicate mantle, the hot water ice "mantle" convects slowly.

The two families of super-Earths have planets that are diverse in size and amount of water. These characteristics depend on the mixture of elements present as the planet forms. From studying the spectra of many stars, we know that the amount of iron and other heavy elements will be different in different planetary systems. We already know that where in the proto-planetary disk a planet forms also matters. So, among the rocky planets we could find super-Mercurys—planets that have as much as 70 percent of iron core inside, like our planet Mercury.[17] Or we could find super-Moons—planets that have no iron core, just an iron-rich mantle and perhaps a water layer.[18]

There is a third possible family of super-Earths and terrestrial planets—carbon planets. These would be extremely rare, as they require more carbon than oxygen to be present in the planet-forming mixture.[19] Normally carbon is half as abundant as oxygen, as we saw in Figure 2.1. But astronomers have observed rare stars in which carbon is more plentiful than oxygen. A planet that forms from such a mixture will be different—it will have a mantle rich in silicon carbide and graphite in its interior.[20] It will still have a precipitated iron core, but its overall size and the chemistry on its surface and crust will be very different. Silicon carbide is a very hardy substance—we use it to make durable ceramics, the disk brakes of sports cars, and tools for other high-stress environments. So volcanism, tectonics, and weathering are going to be minimal on carbon planets. Also, carbon planets are likely deficient in water.

Figure 5.2 shows some imagined family portraits of super-Earths and compares them to Earth and Neptune. The relative sizes of the super-Earths are accurate to the best of our theoretical models. Carbon planets are not shown in Figure 5.2. They will have intermediate sizes between rocky and ocean planets. The Neptune-like giant planet orbiting the star Gliese 436 is shown for comparison too.[21]

Finally, in the approximate mass range for super-Earths we discover small planets with relatively large amounts of hydrogen and helium—perhaps up to 10 percent by mass—just as we see on Neptune. Over all, of course, the planets would be smaller than Neptune: call them mini-Neptunes. It is still unknown where the transition occurs from planets rich in solids to planets with increasingly more massive hydrogen-helium gas envelopes. Theory gives us multiple possible solutions and no clear choices, so we'll need the observations to show us nature's preferences, and NASA's Kepler is well on its way to provide them. Fortunately, it appears that studying the colors of such planets (a.k.a. spectroscopy) will allow separating the mini-Neptunes from the super-Earths.[22]

Even this doesn't exhaust all the possibilities of terrestrial planets in the Universe. Another possible planet would simply be a bare iron core! Under most conditions, we would not expect to see such a thing because a planet body has to be assembled first (at which point it would mostly be silicate, as we've seen) and then differentiate, and only then have the

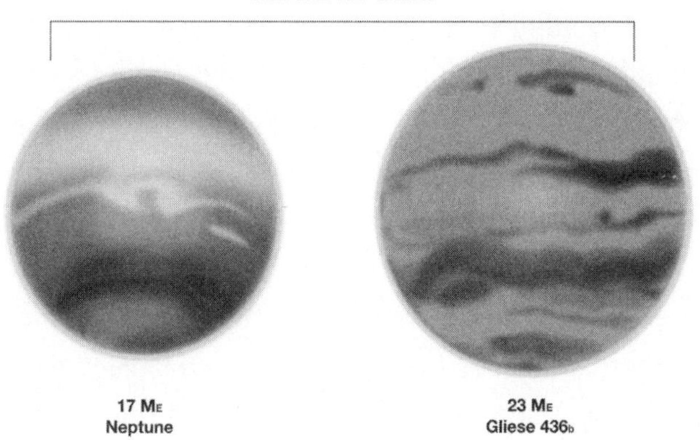


**FIGURE 5.2.** Comparison of planet sizes among different super-Earths, as well as giants.

iron precipitate in a core. The silicate mantle, without which there would be no iron core, would have to be stripped away to leave behind a bare iron core. That's easy to do to a small body—a twelve-mile-wide asteroid, for example. Super-Earths are a different matter; they are so massive that even two super-Earths colliding head-on would not shatter to bits and expose blobs of pure iron.[23] Even under such stress, their gravitational reach stretches far beyond their surface and brings rocks back in. In fact, our own planet Mercury, as small as it is compared to Earth (just 1/18 of Earth's mass), is reported to have survived such a head-on collision early in its history and still retains most of its mantle. Nevertheless, iron super-Earths probably do exist, and we may have already discovered them—the pulsar planets.

Pulsar planets were discovered by Alex Wolszczan and Dale Frail in 1992, marking the dawn of the hunt for exoplanets. As the name suggests, these planets orbit a pulsar—an ultradense, very massive (more than our Sun), fast spinning neutron star; a remnant from the explosion of a spent massive star, known as a supernova. We do not know their sizes (and there is no obvious way to measure them in the near future), nor do we have a good idea how they could have formed. Pulsar planets appear to have accumulated from the iron-rich debris left after the supernova explosion. Heavy-metal planets!

There are limits, however. We wouldn't expect, for example, to see a pure-water planet. Even if we allow for the water to be mixed with ammonia and a small percentage of other impurities, a pure super-Snowball is as unlikely as a snowball on Venus. The problem, as with the pure iron super-Earths, lies again in planet formation. There is no conceivable way to "purify" the snowflakes from the dust in the debris disk or to shield the water-rich planet from accreting rocky bodies. Collisions don't seem to help either, and for the same reason—super-Earths are just too massive. Even in our Solar System, the most water-rich bodies are comets and distant cousins of Pluto with water-rock ratios of about 1:1. If Kepler discovers something with a mean density of water, I would bet on it being an extraterrestrial civilization's container for water storage . . . or perhaps just a mixture of hydrogen gas, water, and rocks—a mini-Neptune planet.

Now that we've considered this range of planets, you may be wondering if it's fair to use the name of our beautiful planet Earth for such a plethora of exotic worlds, even if the models of these other planets are fairly based on knowledge of our own. To me, the answer is clearly yes. Our Earth is a super-Earth, part of the big family, and something general and deep unites our planet with those others. It's so deep, in fact, that you'll find it at the bottom of Earth's mantle. Let's see what it is.

CHAPTER SIX

# SUPER-EARTHS

## *The Hardest Rocks in the Universe*

Rock hunting is fun: get yourself a hammer, a chisel, and a guidebook and head to the hills. If you're lucky (and persistent), you'll discover a pretty specimen to join your collection on the mantle or the bookshelf at home. There are so many—a seemingly endless variety. And the average rock hunter is just scratching the surface of the planet. If only we could look inside it, what wonders might we find? Actually, if you could go deeper inside the Earth, the variety of minerals would decrease dramatically to just a handful, with one of them—perovskite—dominating the bulk of the body of our planet (40 percent of its mass).

Perovskite, despite its planetary abundance, was not noticed and classified by geologists until 1837. The discovery was set in motion in the 1820s, when Czar Nicolas I of Russia, eager to map the uncharted riches of Siberia, invited the accomplished German naturalist Alexander von Humboldt to lead an expedition into the region. Humboldt set out east in the spring of 1829, taking two associates with him. They explored the Ural Mountains and the Caspian Sea, and reached the Altai Mountains of central Asia, where China, Russia, and Mongolia converge.

Reportedly Humboldt did not enjoy the trip much, though it appears to have been very successful.[1] One of his associates, Gustav Rose, was a geologist. Among the minerals he brought back was a heavy dark piece found in the Ural region near the town of Slatoust. It had not been catalogued, so Rose had naming rights as its discoverer. He named it perovskite (originally, peroffskite), in honor of the Russian geologist Count Lev Alexeevich von Perovski, who was also the expedition's host in Moscow.[2] Gustav Rose had no idea that he had discovered the most abundant mineral on Earth!

Rocks are aggregates of one or more minerals. Minerals are inorganic solids with a definite chemical composition. They range from a single element (e.g., gold) to very complex silicates. On planets the predominant minerals are oxides—one or more oxygen atoms bound to silicon, iron, magnesium, aluminum, and so on. For example, silica, also known as quartz, consists of one silicon atom and two oxygen atoms. In silicate minerals that are most common in rocks, one silicon atom is usually surrounded by several oxygen atoms. Depending on how the atoms arrange themselves in regular repeating patterns, different crystal structures are possible.

The perovskite mineral Gustav Rose discovered in Russia was an oxide of titanium and calcium. There is a rich group (known as the perovskite group) of minerals that all share the same unique crystal structure of connected octahedrons of silicon or titanium bound to six oxygen atoms.[3] Despite the fact that Rose discovered a titanium perovskite first, Earth's lower mantle is, by volume, almost entirely taken up by silicon perovskite minerals.

Perovskite minerals contain rare earth metals (such as lanthanum, neodymium, and niobium) as trace elements. Perovskite is famous to physicists as the mineral in which high-temperature superconductivity was first discovered in 1986 by Alex Muller and Georg Bednorz.[4] Nevertheless, most people have never heard of it, and it rarely appears in the popular field guides to rocks and minerals.[5] (More common is a mineral called enstatite, similar to perovskite, which can be found in a variety of rocks.) Importance need not bring fame.

It turns out that perovskite—specifically as it behaves under high pressure—is very important to the study of super-Earths, and indeed forms the link between our planet and other planets. The breakthrough occurred in 2004, when a long-standing mystery about the Earth's interior was finally explained. Seismologists and other scientists had noticed a thin layer at the bottom of the mantle, just above the core-mantle boundary, that had an unpredicted effect on the behavior of earthquake waves propagating through it. This layer, known as D", is only 150 kilometers thick, and no one knew how to model it or why it affected seismic waves as it did. A combination of high-pressure experiments by Murukami and team and theoretical calculations by Oganov and Ono uncovered the culprit—perovskite under very high pressure.[6] At that depth—where the pressure reaches 1.3 million atmospheres—the closely packed structure of perovskite deforms into a layered structure of nanoscale sheets—a new phase the discoverers called *post-perovskite*.[7]

As already noted, this layer is very small on our planet—just 150 kilometers—but turns out to be the main component of a super-Earth! For any super-Earth of about 2 $M_E$ and above, the pressure inside most of the mantle exceeds 1.3 million atmospheres; therefore, super-Earths, whether rocky or oceanic, are post-perovskite planets.

You might have noticed the unremarkable naming of the new mineral. No naming rights here. This is because post-perovskite cannot be present in a rock on the surface of the Earth, so no one has held it in their hands yet, or ever will. (This is apparently what it takes to name a mineral according to the medieval rules of the geological societies.) This is because the structure of post-perovskite makes it act like a loaded spring; if it were ever brought nearer to the surface by convection, it would revert back to perovskite.

Discovering post-perovskite was timely for the upcoming study of super-Earths, but we might not be finished with the project yet. The problem is that pressure throughout the mantle of the most massive super-Earths reaches 10 million atmospheres, much higher than what is needed to turn perovskite into a post-perovskite.[8] It is possible that other mineral phases might be inside them—they could even be a *post*-post-perovskite?

Theory and experiment have already given some hints to the fate of post-perovskite under higher pressure. The main player in this high-pressure drama is the oxygen atom. After submitting to two dense crystalline arrangements, called packing, the further belt tightening leads the oxygen atoms to rearrange themselves again. Their efficient packing seems to have one more possible crystalline structure—a 12-point phase.

The hints come from the study of a little-known crystal called gadolinium gallium garnet—GGG ($Gd_3Ga_5O_{12}$). GGG is transparent and very hard under pressure—useful properties to the scientists who try to compress hydrogen to a million atmospheres in the hope of making thermonuclear energy commercially viable.[9] The compression of hydrogen (e.g., very small pellets of it) is done by powerful lasers, happens quickly, lasts briefly, and has to be observed through a tiny window that does not break under high pressure. In 2005 scientists in Japan and the United States used strong laser impulses to squeeze GGG to 1.2 million atmospheres and found, somewhat to their surprise, that the crystal deformed into a state that was harder (less compressible) than diamond.[10] This was a world record. So, GGG emerged as a better material for the high-pressure experiments with hydrogen, replacing the sapphire and diamond that had provided the tiny windows until then.

The GGG experiment shows that an even denser packing of the oxygen atoms is possible in a post-perovskite material. It may be the last such deformation we need to consider inside a super-Earth planet. Fortunately we can do the experiment in a lab, although it is expensive, so the wait for the answer should not be long. In the meantime, it is already clear that super-Earths are made of the hardest materials known in the Universe: post-perovskite in the rocky and water ones, and diamond in the rare carbon planets. Indeed, this captures part of the beauty of studying super-Earths: the richness of new exotic materials they bring to the table, or lab bench as it may be. Hot water ices VII, X, and XI are as unusual to our world and the rest of the Universe, as are post-perovskite and GGG. Under pressure the solid form of water attains the more compact cubic crystal structure. Ice VII is analogous to crystal rock salt in that way and happens to have similar mean density. What a strange world!

# PART II

# ORIGINS OF LIFE

CHAPTER SEVEN

# THE SCALE OF LIFE

Shakespeare once said that all the world's a stage, and all the men and women merely players.[1] To paraphrase Shakespeare, scientists are acting in a play not knowing the script (or if one exists) and not knowing if there is anyone in the audience. And the stage is certainly not built to our size. It is way too big, even by the standards of fairy tales. Yet we humans are funny little creatures—we possess reason and a spirit capable of matching the vastness of the world.

When I was in high school, I had a small telescope. My father had shown me how to use it, mostly during the day; I was on my own at night, spending hours out in the backyard. The place where I grew up was small, no telescope or observatory at school, no planetarium to visit nearby. So I had almost no idea what I would see through my small telescope. Sure, I had read books, some with pictures, but the view through the telescope was something else. The experience was truly visceral—shivers would run up and down my spine every time I pointed the telescope at a patch of stars. Eventually the feeling went away, but I still remember it. As I looked at the darkness between the stars, I felt as if I could fall into it. Like a fear of heights in reverse. Like an upside-down vertigo.

Perhaps I had read one book too many, and I was imagining things too vividly. After all, I had read about the vastness of the empty space between the stars in my books. But if we had a feel for how big the Universe is, we would have permanent shivers up our spines. To stay sane, astronomers use math, lots of it, and this can ruin even the best party.* Seriously, though, this is a way to deal with the problem. Since the time of Eratosthenes in ancient Greece, who measured the size of the Earth, humans have used math (and geometry) to take the measure of the Universe. Almost always the new knowledge increases our feeling of wonder. Many scientists (and I am one of them) will tell you that this is why they do what they do.

Ironically, this vastness—as unfit as it may be to our tiny scale**—may be essential for life to emerge and survive. So

---

*Walt Whitman's poem "When I Heard the Learn'd Astronomer" has upset generations of scientists, as they will tell you that math does not keep them from looking up in awe at the stars.

**Note the new term I have introduced: "scale." In astrophysics this term means a characteristic range—of size, or of time, energy, and so on. For example, we speak of a galactic scale (sizes of $10^{20}$ to $10^{25}$ m)

let us explore it some. I will use the word "scale" often from now on; it means the extent or relative size of something, whether space or time. When it applies to time, I use "timescale." Let's deal with space first.

We live in a galaxy—the Milky Way—an "island" of stars and gas swirling in spiral arms around a center. The Universe, as seen through telescopes, is filled with galaxies. The current estimate is that there are at least 200 billion of them.

My night sky exploits as a high school youth included viewing different galaxies. It was a challenge, since most galaxies require dark, clear skies and a sizable telescope. However, there is one galaxy that can be seen with an unaided eye, and, unlike most galaxies, it even has a name: Andromeda. If you live north of the equator, you can try to see it—a faint nebulous smudge in the constellation of Andromeda—rising in the east during the late summer nights, and overhead during the fall and winter evenings. I recommend you try hard because this is—by far—the most distant object a human can ever see unaided, with no help from telescopes or any technology. The Andromeda galaxy is 2.5 million light-years away, about 10,000 times farther than the average stars you see at night.

The Andromeda galaxy is very similar to our own Milky Way. Andromeda is a similar flattened disk of stars and gas, most of them bunched in spiral arms, and it's roughly the

---

versus a quantum scale (sizes of $10^{-10}$ to $10^{-15}$ m); and we use shorthand notation—powers of ten—to write those very large (or very small) numbers.

same size. If you are blessed with a dark, clear sky, you will notice that the "smudge" of the Andromeda galaxy appears to be an elongated ellipse. This is because its disk is oriented sideways to us.

Galaxies are mostly close to each other; you could build a scale model of our Milky Way Galaxy neighborhood in your living room. If we were to take our Galaxy to be a dinner plate, then the Andromeda galaxy would be a dinner plate about twelve feet away, and the Triangulum galaxy (another neighboring galaxy, known as M33) would be a salad plate about ten feet away from the Milky Way and a bit to the side of Andromeda. A dozen M&Ms could stand in for the multitude of dwarf satellite galaxies. This is common for our Universe. The galaxies that fill it wall to wall are separated from each other by distances that are comparable to their sizes. You can visualize how this would go in all directions. They are just far enough apart not to bother each other too much.

The picture changes dramatically in the world of the stars, and then again in the world of the planets. You can't build a scaled model of the Sun's stellar neighborhood in your living room—the stars are minuscule compared to the distances between them. While for the galaxies, the ratio between size and distance would be about 1:50 to 1:10 (like comparing M&Ms to dinner plates), that same ratio for stars would be 1:100,000,000 and more (like comparing humans to atoms). With planets, it is similarly huge, albeit less so than for stars. Such is the world!

Is this relevant to life?

One answer could be that it is not. These different scales just happened to be what they are, and that's all. Or perhaps not. Life is a system—a chemical system—that, at least as we know it, seems to work only on small scales. We do not know what life is, but we do know what some of its basic functions are. There is something special about the scale occupied by life to ensure a stable environment that allows such functions to develop. Let us try to understand this by returning to the big picture.

Galaxies in the Universe move with respect to each other with speeds of about 500 kilometers per second. Stars in the galaxy move with similar speeds on their orbits and slightly slower (say, 50 to 200 kilometers per second) with respect to each other. Such speeds are mind-boggling for our everyday experience; for example, a bullet is about 100 times slower.

Here is the problem: such speeds are still minuscule for the distances between galaxies. The Andromeda galaxy is approaching ours at 400 kilometers per second but will require 3 billion years to come close (and may in fact collide with us).[2] Not so for stars! At such speeds, if stars had sizes comparable to the distances between them, they would be running into each other all the time—not to speak of the fate of any orbiting planets. Fortunately, stars don't exist on such scales, so collisions between them are exceedingly rare. Even if the Andromeda galaxy smashes into the Milky Way in 3 billion years, the stars will not collide. Andromeda stars will just glide past Milky Way stars, and then all will mix and merge their orbits around a common new galaxy.

So, on a galactic scale, there is a relative stability, which is important for life. But how much stability is enough? After all, what is stable enough for a microbe might be chaos and doom for a dinosaur.

This issue is similar to the famous question Erwin Schroedinger asked in 1944, Why is life so big compared with an atom?[3] I ask the question in reverse: Why is life so tiny compared to a planet? To answer his question Schroedinger first pointed out how the basic units of life are large chemical complexes of atoms—large molecules. Large molecules and chemical reactions between them are at the heart of every process associated with life. They store and release energy, carry information that can be inherited, and assemble into filaments, walls, structures, and more.

Schroedinger also pointed out that the small scale of the atoms—a world described by the rules of quantum mechanics—is ever changing and not strictly predictable (quite chaotic, indeed).[4]

He should know, being one of the giants of science who helped develop quantum mechanics, demonstrating how very different it is from the classical mechanics developed three centuries earlier by Isaac Newton. Classical mechanics provides the rules for the large scale and large objects—stars and planets, their orbits, bridges, car engines, and so on. Life is large enough to fit in the realm of classical mechanics, and so too are its essential basic units, the large molecules—but only just so.

In answering his question, Schroedinger suggested that the molecules of life and the cells they build are just large enough to avoid the unpredictable and destructive vagaries of the scale of the atoms—the world of quantum physics. At the same time, life benefits from the richness of chemical bonds that is the hallmark of the atomic scale. From my point of view, the complex molecules and chemical networks of life avoid the violent destructiveness of the very large Universe by inhabiting a scale small enough to allow for many stable environments.

So, it seems that the scale inhabited by life has some special qualities. We can observe curious things if we take a swift tour through the space scales of the Universe. The galaxies move slowly like giant turtles, the stars inside them buzz around like bees, the planets orbiting the stars move faster still, and so on until we reach the scale of the microworld—the quantum world of atoms and electrons. The smaller the scale, the crazier the world seems to appear. In fact, *it is crazy*, and modern physics has a good explanation for it. To oversimplify it a bit, big things move slowly, small things move faster. Just think of the truck and the motorcycle at the stoplight when the light turns green. Remember that mass and speed combine to give you energy, and energy is conserved. If mass goes up, speed must go down. There is order to all this, after all.

There is more to the scale of life, though, than simply the profusion of stable environments. To understand the special qualities that the scale of life has, one needs to know nonliving matter first.

The atomic scale and the atoms are the basic building blocks of ordinary matter, as the ancient Greeks surmised. We still think of this as true, at least for the pure elements of the chemical table, such as carbon, iron, or gold, although we recognize most ordinary matter as being made of compounds of atoms. What the twentieth century revealed, however, is that ordinary matter is really composed of smaller particles. These are called fundamental or elementary particles and fall into three families of four particles (and four antiparticles) each. Most common and familiar among them to us are the light particles of the first family: the electron, the up quark and down quark, and the tiny electron neutrino.

You and I, our planet, our star, are all made up entirely of them, particularly electrons, up quarks, and down quarks. The two types of quarks make the protons (two up plus one down) and neutrons (two down and one up) that combine to form the nucleus of an atom and thus the chemical identity of a given element. The lighter electrons orbit the nucleus and give atoms the ability to bond together into molecules. The nucleus of the atom is where the mass is; the electrons around it are insignificantly light, but they make chemistry possible. When you cook in your kitchen, you are playing with the electrons—breaking and reforming chemical bonds. If eating results in your gaining weight, it is because you have added more quarks to your body.

This is not yet the whole story. We can't forget about the fundamental forces. One particle can affect another; for example, the positive proton of a hydrogen atom keeps a negative electron in orbit. One piece of matter can influence another piece of matter by means of these forces. There are four fundamental forces—the gravitational force, the electromagnetic force, the strong force, and the weak force. Our daily lives are, for almost all purposes, exclusively affected by the first two forces. Gravity keeps us on the ground (which we experience as weight) and electromagnetism allows us to move around (via friction), among other things. A common feature to all forces is that they can be represented by an associated particle (usually of no mass at rest) which contains the smallest quantum (or "packet") of force. The electromagnetic force particle is called a photon. We experience photons as light or radiant heat, for example, or we use them to broadcast our cell phone conversations. The gravitational force particle is called a graviton. All these particles have wave properties, as do all elementary particles, including electrons, protons, and neutrons.

Now consider the scales that are much larger—galaxies, stars, and planetary systems. This is a world governed solely by the force of gravity. There is no friction or any other manifestation of electromagnetic force strong enough to deflect or slow down stars and planets from their paths, and the chemical bonding essential to life doesn't happen. On the other extreme is the scale that is much smaller than our own—the quantum scale. The force of gravity has no influence here; the individual particles have such little mass that only electromagnetic forces rule. On both the very large and the very small scales, there is no shelter for life. The cosmic scale is awash with radiation and beyond freezing cold, pummeling anything that depends on electromagnetism for cohesion. On the smallest scale, things move so fast and unpredictably that nothing as ordered as life stands a chance. Life on Earth exists at a comfortable scale in between those two—let's call it the *large-molecule scale*. Its range starts a few steps above the quantum scale ($10^{-9}$ m) and ends closer to home ($10^{-5}$ m).

The large-molecule scale is the true scale of life as we know it. All essential life processes, information-carrying molecules, and most organisms (the microbes) fit nicely in it (see Figure 7.1). The scale of life itself fits nicely on a planet. And never mind the big plants and animals that have outgrown the large-molecule scale—they are a recent development.

What is special about the large-molecule scale? Gravity is still weak, but the mass of large molecules is no longer negligible, so they respond to the force in measurable ways. An important effect is that the water solutions in which such molecules and their structures exist and function are likely to be affected by gravity. Gravity acts as a stabilizing agent at these scales—counteracting electromagnetic forces and providing equilibrium. Planet Earth is a good example of such balance: it is massive enough to shrink and compress under its own gravity. Its rocks, as we've seen, are squeezed under immense pressure, packing their oxygen, silicon, and iron atoms close together. The electromagnetic forces between these atoms, being repulsive, put a limit on how compressed gravity can make the matter; thus Earth has been stable in this state for billions of years and will continue to be for a long time to come. The same balance keeps our Sun stable and shining over billions of years too.

Smaller things—our bodies, for example—are held likewise together in a pressure balance, but it is not the balance between gravity and electromagnetic forces. (Our bodies have

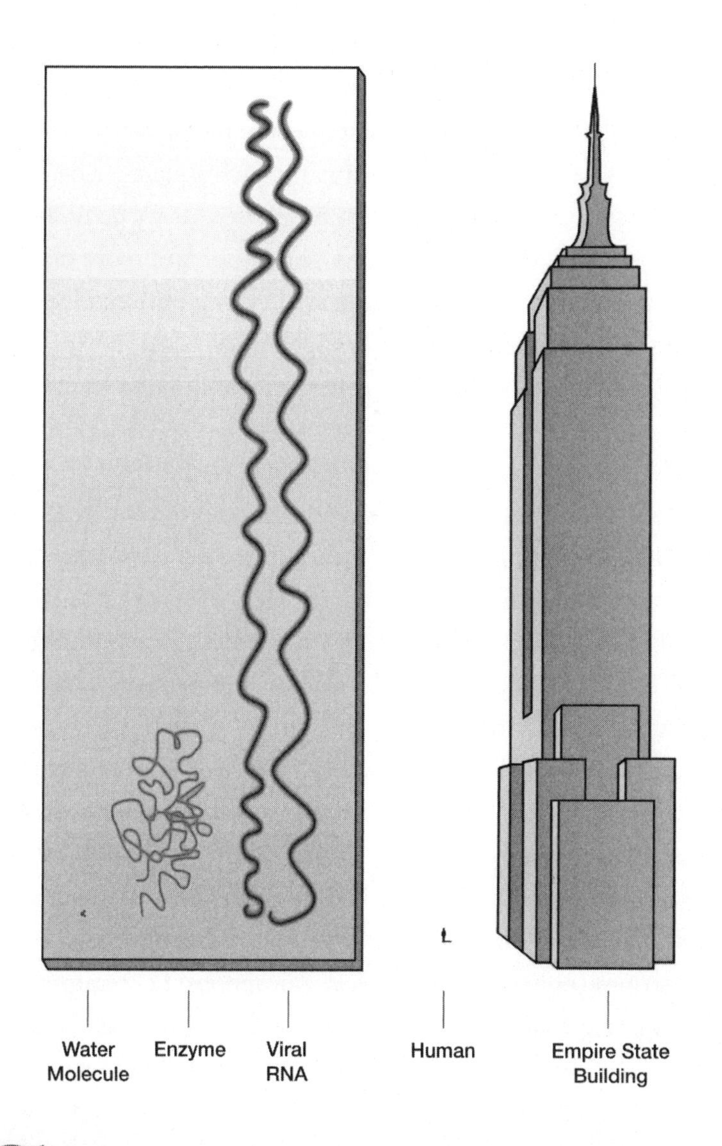


FIGURE 7.1. Comparative sizes: the molecules of life are huge when compared to the common molecule of water ($H_2O$); they define the large-molecule scale.

too little mass for our own gravity to help hold us together.) Instead, we are pulled down by the gravity of planet Earth and simultaneously we are under pressure by the air above our heads—about fifteen pounds of it on each square inch. The air, of course, is also pulled down by gravity. As we go to even smaller scales, electromagnetism gets more and more dominant and gravity all but drops out of the competition. Electromagnetism, as manifested by the chemical bond, makes our tissues structurally sound, keeping cells and larger multicell bodies together.

To everything, then, it seems there is the right size, as J. B. S. Haldane—an early geneticist and famous science popularizer—mused in a 1928 essay. He was answering the question we always ask as children: How are insects built to be able to remain unscathed after a fall from a height that is multiple times their size? Why do some insects walk on water while others drown in it?

Haldane gave the right answers, but only in a general sense. A former colleague, Bill Press, was more mathematical about it and got the correct answer for the size of the human body. To get the correct answer, he had to make three assumptions: that a human body (1) is made of complicated molecules that are strings (polymers) rather than amorphous, (2) requires an atmosphere that is not of hydrogen and helium, and (3) is as large as possible to carry its large brain, but liable to stumble and fall, and in so doing should not break.[5] Thus he defined the right size of an active animal living on the solid surface of planet Earth. That includes the ambient temperature at which humans (and most animals) live, which—not surprisingly—is very close to the temperature of their chemical bond energy. As Bill points out, that fact makes cooking practicable.

The big Universe, however, is definitely not built to our human scale, and the problem is not only the vastness of space, but also the unimaginable scale of time. Time and space are inseparably linked, as we learned from twentieth-century physics, so it should not be surprising that in a big place things take a long time to change. It is essential to my story that you get a feel for this. Nevertheless, while the Universe as a whole is not built to a human scale, it is obviously possible to find places that are, for example, our home planet. This, as I will argue in the next chapter, is not an accident. Life needs planets. Let's see why.

CHAPTER EIGHT

# ORIGINS OF LIFE

## *Why Planets?*

During the Middle Ages, when every continent was a separate world, people seriously considered multiple origins.[1] During the Renaissance, with Earth now just another planet, people seriously considered multiple origins of life on the other planets, as well as the Moon. Thus I am bringing up an old question and revisiting it with fresh evidence about newly discovered worlds and about what it takes for life to emerge and survive on them. Some of this evidence I have touched on earlier, but I am going to bring together the various threads here to argue that for life to emerge, it must happen on planets.

So, let's take a step back and think of the cosmic perspective. Why planets? Aren't there other places in the Universe that could be equally good cradles for life?

Here is a recipe to answer these questions: get what we know about other places in the Universe and mix it with what we know about life, and see if anything useful comes out. The first ingredient is easy to procure. During the past fifty years astronomers have revolutionized our knowledge of the observable Universe and have a pretty complete census of what is in it—lots of galaxies, many more stars, and a list of planets that grows practically by the day.

What we know about life is the difficult part—we know a lot and very little, at the same time. Scientists have revolutionized our understanding of the building blocks and the amazing interconnectedness of living forms, yet the parts are in a system,* they form networks,** and there is something essential but elusive in all that.

I'll begin with the difficult part and list what we know about life—not a definition of life, but just some essential attributes:[2]

1. Life is chemical in essence.[3]
2. Life is a system that is not in equilibrium.[4]
3. Life is adaptive and self-optimizing.[5]
4. Life is compartmentalized—it needs cells, enclosures, vesicles.
5. Life uses molecules that are suited to water.

*The definition of system as a combination of related elements organized into a complex whole is what I use here to refer to life for a lack of a better word.

**A network is a pattern of branching and interconnected lines, circuits, things, and so on; the pattern does not have to be ordered.

The list is easy enough to compile; the difficult part is discerning what may be missing from it, since all we know about life comes from a single example—life on Earth. Curiously, each one of the five items taken separately can describe a nonliving system; life seems to need all five together.

The first attribute is obvious but very useful to the big picture because it is an absolute for any origin of life in our Universe. There is no other form of matter or system capable of ordered networks; chemical bonds can do wonders under natural conditions. We cannot exclude the possibility that extant life-forms might develop a system (or develop *into* a system) that is not based on molecules, but that is not the question I am asking. I am interested in the process by which life emerges and the environments that allow it; that is, I am interested in the cosmic transition from chemistry to life. People have described nonmolecular (nonbiological) life-forms (in, e.g., the writings of Hans Moravec, Ray Kurzweil, and Steven Dick), but always as derivative or evolving from biological predecessors and based, at least partially, on molecules and molecular bonds.[6] We humans may be capable of creating ordered networks in silicon, but this is the result of a long history of the development of technology, not of a lifeless planet's original condition.

Planets help make this possible by providing a physical screen against space radiation, through their bulk and their atmosphere. Planets provide stability over an average timescale longer than the development of chemical networks—years. For example, interstellar molecular clouds (like any cloud) cannot provide a macro- or microenvironment that is stable or protected from destructive cosmic radiation on such timescales, while planets can.

The second attribute is also an absolute: a system at equilibrium is a dead system in which nothing happens; you need energy to accomplish anything. To assemble an ordered network—chemical or not—you need energy. You need energy just to keep it ordered, to keep the inside separate from the outside.

The third attribute is the most fascinating thing about Earth life, and probably of any life—it can adapt itself to both fast and slow changes. Charles Darwin's genius was to see the essence of that property, which is based in part on the process by which new and different organisms develop as a result of changes in genetic material. We call it evolution, or Darwinian evolution in his honor. Most scientists consider it so important that they would define life through it. Gerald Joyce of the Scripps Institute summarized a NASA committee proposal in 1994 in a famous short definition of life as "a self-sustaining chemical system capable of Darwinian evolution," sometimes referred to as the "NASA definition."[7] Even though Darwinian evolution is an obvious way[8] to achieve diversity in response to changing environmental conditions and interaction with other living forms,[9] strictly speaking, it is not the only one.

The alternative is the existence of environmental conditions that allow continuous creation of life-forms with no inheritable molecules, but still with random variations.[10] The burden to sustain such a biochemistry then rests on the existence of environmental conditions that allow the continuous synthesis of life-forms. Evolution offers a simpler and more straightforward mechanism, and is very likely a universal attribute to different origins of life and different biochemistries.

In the cosmic transition from chemistry to life there is no substitute for molecules and what they can do; only one of the essential attributes—the second, describing energy dissipation—can be worked out without molecules. Any of the others will be impossible to do without some sort of complex molecules—usually large ones, called polymers. Polymers are very long molecules made of smaller units called monomers (see Figure 7.1). The basic polymers of life, called biopolymers, are proteins—made by linking together amino acids and nucleic acids—made from nucleotides.

We might not know which chemical networks and attributes are essential to life, but as long as we are certain that some of them are, then molecules become building blocks and we can't do without them. Moreover, even though there are millions of molecules, the toolbox for building them is very limited, so we can identify some fairly narrow constraints on what must be around for life to happen. This is a good thing, as it gives us a fighting chance to understand them.

As we've seen, life must exist on a scale at which both gravity and electromagnetism are noticeable forces, and life needs protection from the environment of space. Consider temperature. In the Universe there is a wide range—from a few degrees above absolute zero to several hundred million degrees.[11] The coldest places are clouds of gas away from stars in the outskirts of small galaxies. Many planets are very cold too. In our Solar System the dwarf planets Sedna and Eris are more than three times farther from the Sun than Neptune, and their surface temperature never rises above a bone-chilling 30 K (–400°F). The gas giant planets that were photographed recently around stars HR 8799 and Fomalhaut are at similarly great distances from their parent stars. It is expected that many more planets are equally cold.

At the hot extreme of temperature are the envelopes of massive stars and gas near the central regions of galaxies. We know the destruction tolerances of molecules, and most of the high-temperature range is off-limits. While a few very hardy ones, such as carbon monoxide, survive at high stellar temperature (in low densities, they can survive temperatures as high as 6,000 K), any molecule with more than three atoms needs temperatures lower than those found in stars. Gas near the central regions of galaxies won't work either. The X-ray and UV light emitted by stars, as well as the energy released during the explosions of supernovae, keep lots of the gas in galaxies hot too.

The very low temperatures have their problems too—polymers survive but lose chemical functionality. Just as we know from personal experience (and make use of, through refrigeration), chemistry happens very slowly if at all at very low temperature. So, there is a narrow range of temperatures in which large molecules thrive and complex networks of chemical reactions can take place. That temperature range is from about 100 K to 600 K if we feel generous, but much narrower if we consider Earth biochemistry as we know it today.

That is a remarkable conclusion. In a Universe where temperatures of millions of degrees are common from a cosmic perspective, life is a low-temperature phenomenon. At these low temperatures there are only a couple of large long-lasting types of objects in our Universe—clouds of gas between the stars, called molecular clouds, and planets (as well as their satellites and other small planetary bodies).

Access to energy is another important environmental condition. In the above temperature range, there are two benign, steady, and long-lasting sources of energy in moderate amounts: starlight and internal heat of planets. Being a certain distance away from a star, such as Earth is from the Sun, is an excellent place. The energy flow is steady and moderate, meaning that it matches the energy needs of large molecules without destroying them readily. To access this stellar energy, you do not have to be on a planet. On the other hand, planetary hydrothermal vents, such as those in Yellowstone National Park or Iceland, or on the ocean floor, are a good example of how internal heat can provide a similar source of energy, both locally and globally on a planet, no matter how far from the star. Planets cool slowly, especially larger super-Earths, and this source can be steady for a very long time. In addition to heat, vents provide a rich source of chemicals that can be used as a source of energy. Hydrogen sulfide, for example, a chemical poisonous to you and me, is harnessed by bacteria at vents through a process known as chemosynthesis to power their cellular machinery.

Speaking of chemicals, there is a third important environmental condition—life seems to need environments allowing chemical concentration. Most environments in the Universe are dilute. Relatively complex molecules are observed to form in molecular clouds, but their concentrations are always extremely dilute. This severely limits the complexity in molecules and their reaction networks. Gas giant planets like Jupiter or Neptune are another example of chemically dilute places. Their cloudy atmospheres and bright display of colors are all there is—no solid or liquid surface. As you plunge deeper below the clouds, the gas gets denser and hotter, and all but the smallest molecules are destroyed. Life on Earth uses enclosures such as cells and vesicles within them to further concentrate chemicals in useful places; maintaining this kind of disequilibrium, of course, requires energy.

The combination of relatively low temperatures, steady sources of energy, and access to chemical concentration leaves us with planets as the best, if not the only, places where the molecules of life and their interrelated reactions can emerge and sustain themselves. No other type of object has the complete set of conditions. And then only some planets—gas giants won't do. Terrestrial planets are unique in providing a range of rich chemical concentrations, energy sources, and sheltered environments. This is a profound realization!

CHAPTER NINE

# LIFE AS A PLANETARY PHENOMENON

Half a day's sail northwest of the Cape of Good Hope lies Cape Town, South Africa. Even today, the city feels like an outpost, a cozy yet uneasy harbor at the edge of the world. After all, there is, to human eyes, nothing south of there but cold and ice, a frigid southern ocean encircling the least hospitable landmass on Earth.

There, in the shadow of exotic Table Mountain, in November 1873 an unusually outfitted British ship, HMS *Challenger*, was being readied for a grueling journey to the icy shores of Antarctica, then on to Australia and around the world. Forty years after the famous trip of the HMS *Beagle* with Charles Darwin onboard, HMS *Challenger* was on a four-year voyage to explore a new frontier—the depths of the world oceans.[1] HMS *Challenger* had a complement of six scientists in its crew and a deck loaded with curious instruments: for measuring oceanic temperatures, depths, and currents; for taking deep-ocean samples; and for capturing living specimens of whatever creatures might call the strange place home.[2]

The results astounded laypeople and scientists alike. From the total darkness and tremendous pressure found eight or more kilometers below the surface, HMS *Challenger* collected 600 cases of specimens. The rich harvest proved that the vast underwater landscape of the oceans is not a desert. Life was everywhere on the surface of our planet.

More recently, research in the last twenty years has led to two more astonishing realizations. First, rather than simply providing a home for life, planet Earth has been thoroughly transformed by life, which has accompanied it almost since its formation 4.5 billion years ago. What's more, as fragile as life may appear to be, clinging to the surface of a small planet that is subject to violent cosmic events, what we have learned has led us to conclude that life on Earth is virtually indestructible. The evidence suggests that it has been that way for a very long time, perhaps 3 billion years.[3] Destroying all living things on Earth, including spores and complex biomolecules, would take nothing short of melting and vaporizing the planet in the sterile interior of the Sun. Destroying life would also require annihilating all the spacecraft in Earth orbit, as well as the other places we've sent space probes. Though unlikely, the possibility that we've colonized Mars with some microbes, for example, is not that far-fetched; many extremely resilient microbes have been discovered in the past thirty years, and some—such as *Deinococcus radiodurans*—are hardy enough to survive the trip to Mars.[4] Real space travelers!

The secret behind the indestructibility of the Earth biosphere lies in the sheer diversity and inventiveness of the organisms that have always ruled "our" planet—the microbes. The hardiest among them are called extremophiles, meaning that they inhabit extreme environments. Some are able to withstand 250°F (122°C) in hot springs and in ocean floor "black smokers"—hot volcanic vents. Others survive high pressure levels, for example, in the sterilizing high pressure vats for orange juice[5] or in the natural habitat of the Marianas trench in the Pacific Ocean at 800 atmospheres.[6] Yet others have made their home in the tiny cracks of rocks four kilometers underground, discovered as history has gone through another cycle, and a flotilla of twenty-first-century ships, mimicking the *Challenger* before them, has drilled into the ocean floor.[7]

Scientists have probably not yet reached the bottom of the Earth's biosphere. Microbes that live deep within the Earth's crust are often dubbed SLiME communities, for subsurface lithoauthotrophic microbial ecosystem. They depend on nothing from the surface. The heat comes from the depths of the planet, the chemicals and water are already there, and sunlight is not needed. Though extreme, their environment is very protected.

These are not anomalous creatures, either. The "bottom" line is that life on Earth appears indestructible today because this subsurface environment hosts a large fraction of the planet's total life. Some scientists, such as the late Thomas Gold at Cornell University, have argued that indeed most of the biomass on planet Earth is below the surface. Recent estimates are up to 300 billion tons of carbon biomass, which is comparable to the entire continental surface biomass, which is mostly plant life.[8] Most of this deep biosphere consists of microbial communities living in rocks and sediments roughly 500 to 1,000 meters below the ocean floor; a case at 1,600 meters below the Atlantic seabed is the current record holder.[9] With the ocean floor covering more than 70 percent of our planet, and a measured million cells in every cubic centimeter of subfloor sediment, this would make for more than half of all microbial cells on Earth.[10]

Most recently, deep biosphere hunters discovered the first nonmicrobial life-form from a 1- to 3.5-kilometer depth in South Africa—a tiny worm that feeds on subsurface bacteria.[11] This emphasizes the richness of life in the deep crust of the Earth.

How can microbes survive in miles of rock without sunlight or oxygen, and having scarce nutrients and water? The drill samples show a predominance of microbes that are resilient to stress and especially skillful in conserving energy by growing (doubling) extremely slowly—on timescales of centuries! If they had any cares, they would not be about us, the surface dwellers, and yet it is clear that they descended, albeit hundreds of millions of years ago, from ocean floor and surface dwellers. This is revealed in their genome maps. They are not that extraordinary, after all.

The history of life on Earth shows rapid adaptation and colonization of any place where there is water, regardless of extremes in temperature, pressure, and acidity. The deep water cycle—the water from the surface that reaches deep into the crust and below the oceans—has brought life along with it, probably as soon as life existed on this planet.

What dangers there are to life mostly come from outside Earth. The most dramatic threats are cosmic: asteroid and comet collisions, as well as major atmospheric change, including the loss of the entire atmosphere. Dramatic, yes, but not necessarily a death blow to life on the planet.

Asteroid impacts are a part of life in any planetary system. Asteroids, the mile-size (sometimes many miles) fragments left over from the accumulation of the rocky planets, have orbits that are prone to be influenced by the big planets. Many of them, as a result, have been "swept up" by larger planets in the Solar System, including Jupiter, but many remain, especially in a large belt that exists between Mars and Jupiter. Over long periods of time, the gravitational influence of the planets is enough to put the asteroids on a collision course with a planet or another asteroid.

An impact by a two kilometer asteroid would be a catastrophe for humankind, but most of the microbes in the deep biosphere would not even notice the event. It would take an impacting body almost the size of Mercury to destroy Earth's crust and oceans and perhaps sterilize all colonies of microorganisms that are miles below the surface. However, in our Solar System, at the start of the twenty-first century, astronomers have a pretty complete census of asteroids crossing Earth's orbit. We know all bodies larger than two kilometers that could hit us. Astronomers know of no such planet threatening to impact the Earth.

Collisions between asteroids and planets would have been very common during the period of planet formation and about 500 million years after. We know this from observing other solar systems. Were large collisions more common, the amount of small particles lingering among the planetary orbits would be more than enough to notice in our remote census of known nearby planetary systems. Such "debris disks," as they are called, are well-known and easy to detect with modern infrared technology. The Spitzer space telescope, an infrared cousin to the Hubble space telescope, has provided evidence that our Solar System is quite typical in that respect.

What about catastrophic climate change—the total loss of the Earth's atmosphere and the loss or freezing over of any remaining oceans. This could occur due to a massive impact by an asteroid comparable to the Moon or a nearby stellar explosion: a supernova or a gamma ray burst.

Gamma rays are the most energetic electromagnetic waves. Gamma ray bursts are among the most violent events we know and they occur infrequently in any given galaxy. Nevertheless, astronomers detect them often—once a day. This is because of their sheer brightness and the penetration of gamma rays. We are able to see a burst across the entire visible Universe. They emanate from the final explosions of very weighty stars (a special case of a supernova explosion) and the spiraling-in of two neutron stars. Nothing rivals the power of the explosion that brings about a gamma ray burst. However, such explosions are both rare and far apart. At the typical rate and average distance, the only effect we should worry about on Earth is ozone layer loss. In the unlikely event that one occurred within fifty light-years of the Solar System, however, we would be in trouble. The Earth's atmosphere would be completely lost and all life on the surface would become extinct.[12] But not life inside the crust.

A sudden loss of the entire atmosphere would likely deprive the Earth of an atmosphere just temporarily, on geological timescales. Because the internal structure of the Earth is not going to change much at all, the basic plate tectonics and the release of gas from the planet's interior through volcanic activity would continue. Carbon dioxide from volcanoes would replenish the atmosphere, which, because carbon dioxide is a greenhouse gas, would melt the frozen oceans (or whatever was left of them). Even if they melted partially, the evaporation of water into the atmosphere, followed by rain and erosion, would restart the carbonate silicate cycle.

The carbonate silicate cycle is almost identical to what is commonly called the inorganic carbon cycle, or the carbon dioxide cycle. It is a fundamental planetary cycle of the abundant gas carbon dioxide as it rises from the Earth's interior, undergoes transformations in the atmosphere, on land, and in the oceans, and returns back inside at the end of it. The carbonate silicate cycle has a typical timescale—a typical time for changes to take hold. For planet Earth this timescale is about 400,000 years. So, should a gamma ray burst destroy Earth's atmosphere, it might take "just" a few million years for it to return and stabilize, perhaps less. Any microbes that survived deep in the crust—and there should be many—would have ample opportunity to recolonize the Earth's surface. For example, microbial communities discovered in deep drilling in Texas appear to have been cut off from the surface 80 million years ago, much longer than the million years needed to recolonize.

Of course, if such a calamity were to happen, the Earth's atmosphere would be changed dramatically: its two main constituent gases today, nitrogen and oxygen, would be gone and could not be replenished by volcanoes and evaporation from the oceans. Of course, this wouldn't be a big problem for any subterranean microorganisms remigrating to the surface; they have no need for oxygen gas in their present location, and would do just fine on the "new old Earth," as long as some access to sources of nitrogen remain in the crust. They might even put those gases back, as they are byproducts of microbial life, if given another billion years—as they already did on Earth about 2 billion years ago.[13]

It is humans and complex life forms that live a precarious existence subject to the vagaries of cosmic change. Earth life, as represented by its most numerous and ancient forms—the microbes—is permanently entrenched on our little planet, at least until the Sun retires and engulfs it. Think about what would happen if the Earth's orbit became unstable and a near collision with Jupiter were to fling Earth out of the Solar System.[14] Sounds like the end of days, literally, as darkness and deep freeze would cover the surface. However, hydrothermal activity—those same black smokers in the middle of the Atlantic Ocean—would continue without interruption. Much of life that calls black smokers home would survive, and for quite some time—the crust makes an excellent blanket, trapping the remnant heat from Earth's formation, as well as the heat emitted by radioactive decay of elements like uranium, potassium 40, and thorium. In fact, the rate of loss of internal heat on Earth today is measured to be 87 milliwatts per square meter.[15] This is nearly a thousand times weaker than the rate at which a household lightbulb uses energy, and you would have to collect all the internal heat from an area larger than a college classroom to light up just a feeble 25-watt lightbulb. This seems like a drop in the ocean for our energy-hungry twenty-first-century human society, but is entirely sufficient for microbes that live deep in the crust and near hydrothermal vents at the bottom of the ocean. At its present rate of cooling, Earth, even lost in space, could keep its hydrothermal habitats alive for at least 5 billion years.[16]

Earth life is extremely resilient, and has been for most of its history. One reason is its inherent ability to adapt to changing conditions and take advantage of varied, even extreme environments. In doing so, it also modifies and creates new environments, eventually transforming the surface of the whole planet. Planet Earth today is very different from the planet on which life emerged—precisely because life emerged on it.

Moreover, Earth has been harshly inhospitable to the emergence of new forms of life for billions of years, courtesy of current Earth life and the highly oxidizing atmosphere it has created. Life is truly a planetary phenomenon, as my colleague Andrew Knoll showed convincingly: life and planet evolved together, dynamic and inexorably linked.[17] Planets may be good places to sustain life, but we have not yet answered the question, Is Earth the ideal place for life to emerge?

CHAPTER TEN

# PLACES WE COULD CALL HOME

A long time ago, in a galaxy a lot like ours, a star formed. Soon several planets formed too. Three of the planets were close to the star. They weren't very big, and they were almost entirely rocky.

The smallest and the coldest of the three had an atmosphere of carbon dioxide and sulfur dioxide produced by volcanoes. They acted like a greenhouse that kept its surface warm, and so some of the surface water was able to liquefy. Together, the atmosphere and the oceans made the chemistry on the planet's surface varied and fun—and gave life a good chance to emerge. The party ended after a few hundred million years, in geological terms nothing more than a summer vacation in Alaska. The planet, being small, had a hard time holding on to its atmosphere and supporting enough geological activity to replenish it. It just could not keep warm, inside or out.

The other two planets, however, were larger, so they had more water and a more substantial atmosphere, though they too were all rock at heart. One of them, the planet closer to the star of the pair, was—like its smaller cousin—quick to change, although in an opposite sense. Its ocean began to evaporate under the star's glare, and the addition of the water to the atmosphere—where it served as an excellent greenhouse gas—prompted a runaway loss of the rest of the water, leaving nothing but a rocky desert behind.[1]

On both planets, however varied the chemistry might have been initially, the geochemistry became increasingly limited, and gradually settled into an inactive equilibrium. The future development of these planets soon became predictable because it was very simple—described by the basic laws of physics and chemistry. They joined most of the rest of the galaxy—from gas giants to stellar remnants—in a slow, dull process of cooling with no further chemical changes, simply whirling around under the influence of gravity.

The third planet was lucky—large enough and warm enough to replenish and keep an atmosphere and to recycle its surface material, yet not too hot, so that it could keep most of its surface water liquid. The chemistry on this planet was fun and varied too, but even more so than on its little cousin. An existing geochemical cycle of carbon dioxide soon took the direction of planetary development farther and farther away from its geochemical equilibrium. Something fundamentally different was happening. About 2 billion years into this, the atmosphere and the oceans became increasingly richer in the very reactive free oxygen. Real fun for a chemist!

While dull equilibrium ever more firmly gripped the first two planets, the third was becoming even wilder, reacting in complex ways to outside disturbances, such as the asteroid impacts that for a time incessantly bombarded the planet's surface. This went on for about 4 billion years, although the asteroid impacts eventually petered out—until finally this complex chemistry harnessed enough energy locally to eject small pieces of the surface (and itself!) into orbit, eventually to its moon and back, eventually away from the star and into the galaxy.

From the perspective just described, the fate of three planets as viewed from above, from the vantage point of the distant stars, might appear very natural. Just as stars are different, yet any star is much simpler than a planet, planets are different too and some develop a complex chemistry. Life as viewed from the stars would be indistinguishable from any other kind of chemical process, except for the complexity of its outputs. Much as we might speak of a water planet, because its chemistry outputs great oceans, we might, in a galactic sense, speak of a life planet because that's what, given its constituents and its place in a solar system, its chemistry leads to. And just like liquid oceans might be only a part of the history of a type of planet, so it could be with life.

You probably won't be surprised to learn that the three planets I've been describing are planets we know well—Mars, Venus, and Earth. In this context, Earth seems like the place to be. But is Earth the ideal planet for life? I come back to the question I posed in the beginning of the book. Now I'll try to answer it. And the answer is no. Super-Earths are better.

For any planet we find, super-Earth or not, we try to assess habitability. It is a tricky concept. For example, if we define as habitable any place we humans can inhabit without special protection, there will be many places on Earth that fail to qualify. If we allow for areas where we could survive given our best technology, the equatorial regions of Mars would qualify but not a fiery furnace like Venus. Some Earth microbes are so hardy they could survive in the Martian soil. Clearly the habitability of a planet depends on its location: far from the Sun it is too cold, much closer to the Sun is too hot. This range of comfortable distance is called a habitable zone, and is indicated by the presence of surface water.[2] Every star has one.

I am looking for environments that make complex molecular chemistry viable. I am interested in pathways to (origins of) life in the Universe, so I am looking for planets that maintain surface temperatures in which large molecules can survive and can attain chemical concentrations and can be sufficiently stable over time. Although the habitable zone concept is helpful in terms of temperature, it is not sufficient. There are many other factors that contribute to the habitability of a planet. Given that our knowledge of the conditions on super-Earths will be very limited until we actually visit them one day in the future, I prefer to talk about their *habitable potential.*

What is the habitable potential of the first known super-Earths? Let's visit some of them. Gliese 876d orbits a small cool M dwarf star once every two days! That means that its orbit is only 2 percent of the Earth-Sun distance (or 0.02 AU). Its close proximity means that, although its star is only a third of the weight and size of the Sun and only half as hot, the surface temperature on Gliese 876d must exceed that of sweltering Venus (about 730 K). Super-Earth Gliese 876d is not in the habitable zone and has no habitable potential (Figure 10.1).

There are two other, Jupiter-like, planets in this planetary system.[3] They were discovered in 2000 by Geoff Marcy and the California-Carnegie team and have the designations Gliese 876b and 876c. They have masses equal to about 2 and 0.6 Jupiters, respectively. Despite being inside the habitable zone, these planets are gas giants. Lacking any solid surface, they have no habitable potential. However, either of them might have large moons that could have excellent habitable potential. Unfortunately, these two Jupiter-like planets have close orbits and interact gravitationally with each other, so it is likely that they could not have and retain large moons.

Another M dwarf star, Gliese 581, is known to have three super-Earths (Figure 10.1). A hot Neptune of 25 $M_E$ (called Gliese 581b) has been known since 2005; it orbits the star with a period of 5.4 days. Later, Michel Mayor's Geneva team found two smaller planets that could be the closest resemblance to Earth yet, as they seem to be barely inside the habitable zone. Because Gliese 581 and 876 are very similar stars, it is easy to compare their planetary systems and habitable zones. The simple assumption that super-Earths ought to have atmospheres leads to the habitable zone shown in the figure, derived from the work of Franck Selsis and colleagues.[4] The super-Earth with the best habitable potential is the planet Gliese

**Gliese 876**

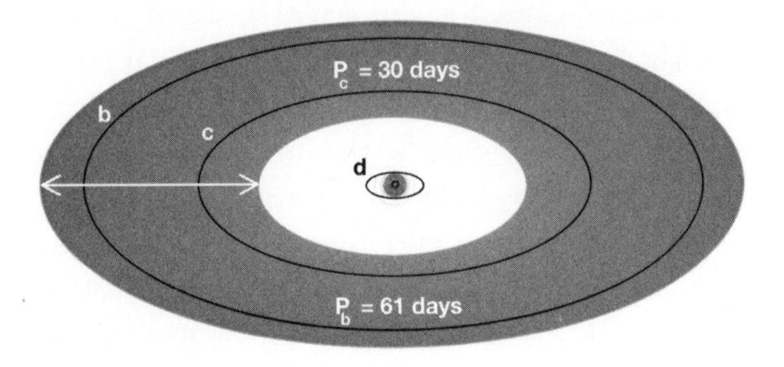


**Gliese 581**

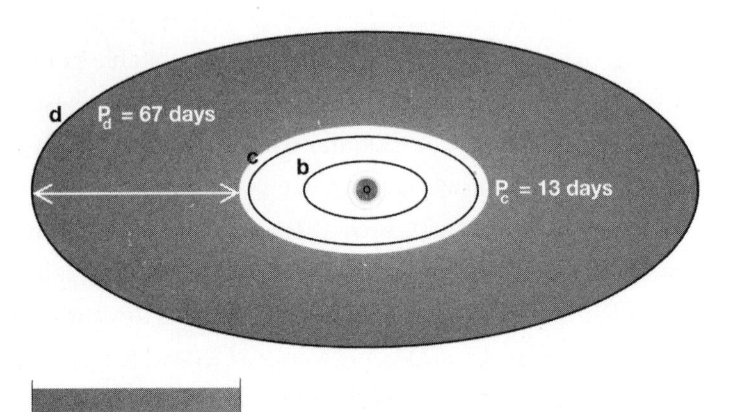


**Habitable Zone**

**FIGURE 10.1.** Two planetary systems and their habitable zones. The super-Earth in Gliese 876 is the inner D planet. In Gliese 581, the inner B planet is actually a hot Neptune. Planet Gliese 581d is the only super-Earth that happens to orbit within a habitable zone, barely; it has a slightly eccentric orbit. Gliese 581e, a fourth planet not shown here, is barely two Earths but orbits the star inside the orbit of planet B and is too hot.

581d, which has a mass of eight Earths and is farther from the star; its atmosphere helps keep the planet warm. Gliese 581c, which has a mass of five Earths, is likely too hot, and there is an even hotter super-Earth E inside the orbit of B.

The planetary system Gliese 581 is interesting in another way: most, if not all, of the planets must have formed farther from the star than they currently are. First, there is not enough material in a proto-planetary disk for such a bulky planet as the hot Neptune, Gliese 581b, to have formed just a dozen stellar sizes away from the star. Second, the temperature in the disk is too high for the large planet to have accumulated. It is much more likely that the hot Neptune, just like the hot Jupiters described earlier, formed farther out and was pushed close to the star by the disk. If that's true, then the two super-Earths must have formed even farther out in the disk. They could be ocean planets, as they must have formed outside the disk's snow line, where water freezes and accumulates easily. Some of this surface water would have liquefied when the planets were pushed closer in. However, they could be gas-rich planets, similar to the recently discovered planets GJ 1214b by David Charbonneau's MEarth Project, and the planets in the Kepler-11 system. These are all of similar masses—a few Earths, and they are all of very low density that requires them to be mini versions of Uranus and Neptune.

Two factors make super-Earth planets more habitable than a planet the size of our own. First, by being more massive these planets have an easier time keeping their atmospheres and water from evaporating. This is very important for the ones that would orbit closer to their stars than, say, Mars is to the Sun. Second, if they are rocky super-Earths, their tectonic plate activity is as high as the Earth's or even higher, according to our theoretical models.[5] This is important for life and its origins. In our Solar System, Mars never had moving plates, and Venus seems marginally capable of moving its plates. It seems that our Earth barely made it!

Tectonic plate activity is what we observe as continental drift on Earth. Modern GPS technology allows us to measure this motion. But why is it beneficial to life? The answer, in a nutshell, is stability and chemical concentration.[6] Over billions of years the Earth has kept its surface temperature stable. The oceans, for example, have generally always been liquid. We know this from geological evidence. We also know that our Sun has brightened up by 30 percent since the Earth formed. The solution to this seeming mystery appears to be a global geochemical cycle. The Earth is a large ball that is very hot and boiling inside. On the surface this energy moves the set of rocky continental and ocean plates around. The essential thing about planetary plate tectonics is *exchange*. The molten, mixed interior can exchange chemical elements with the surface and atmosphere and vice versa. The elements are not simply recycled in this process, but their chemical transformations and concentrations exchange energy in a rich dynamic equilibrium. The alternative is a much poorer steady-state equilibrium that will set in on the inside and at the surface with no local energy sources. Thus plate tectonics makes a planet dynamic, renewing, and vital. As Ward and Brownlee put it, plate tectonics promotes environmental complexity.[7] The exchange occurs through global cycles.

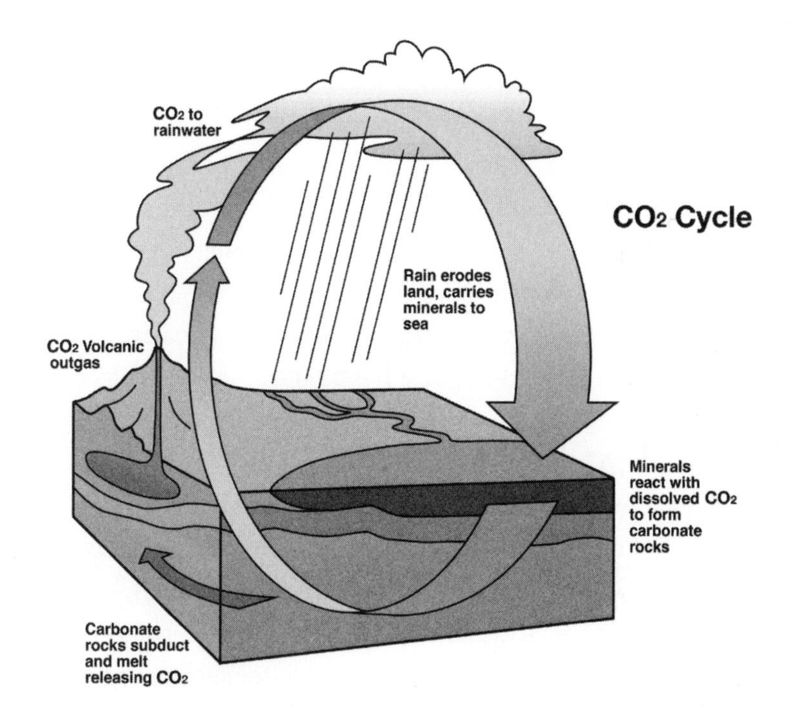


**FIGURE 10.2.** The $CO_2$ cycle (a.k.a. carbonate silicate cycle) starts with the volcanoes on the left that provide the supply of $CO_2$ gas, which gets absorbed by water in the atmosphere, rains down, erodes the continents, and ends up recycled back into the hot interior via plate tectonics, only to return again.

The best-known and dominant global planetary cycle is the carbonate silicate cycle (see Figure 10.2 on preceding page). The cycle begins with $CO_2$ released by volcanoes into the atmosphere, where it is easily absorbed by water droplets. Raining down on the surface, the carbonated water helps erode rocks and soil into the oceans, where it gets deposited in carbon-rich rocks such as limestone. Tectonic plate activity brings the carbon back under the Earth's crust, only to be made molten, mixed, and recycled back into the atmosphere through volcanoes.[8] As we saw in the last chapter, this cycle is important to the existence of our planet's atmosphere, but the cycle does more: it acts like a thermostat, because $CO_2$ is a greenhouse gas and its cycle has a built-in feedback loop that returns Earth's temperature to a normal average (see Figure 10.3 on page 128). A greenhouse gas lets sunlight heat the surface and then helps keep that heat like a blanket. Water vapor and methane gas are other common greenhouse gases. If the Earth's surface temperature increased slightly, the $CO_2$ content of the atmosphere would decrease because the gas would dissolve in plentiful water due to increased evaporation, and rain would carry it to the surface to speed up rock erosion and ocean deposition. This reduced quantity of $CO_2$ in the atmosphere would weaken the greenhouse effect and lower the Earth's surface temperature. As the Earth's surface temperature decreased, the $CO_2$ content of the atmosphere would rise because there would be less rain. In turn, more $CO_2$ would strengthen greenhouse warming and bring the Earth's temperature back to normal. A perfect thermostat!

Well, perfect may be too strong, given the cycles of ice ages that the Earth has endured. The $CO_2$ cycle thermostat has a very long time delay (about 400,000 years).[9] However, ice ages are minor inconveniences in the history of life. Even humans survived the last one 10,000 years ago, and smaller organisms, like subterranean microbes, wouldn't have noticed what was going on. The $CO_2$ cycle has protected our planet from much more serious trouble over billions of years. It will continue to do so as the Sun gets brighter in the next 2–3 billion years. If solar heat continues to rise, a tipping point will come (Venus crossed its own sometime in the past), after which the thermostat will break down.

To the skeptical reader, the $CO_2$ cycle thermostat might seem like a very special feature of our planet Earth. Not so. Carbon and oxygen are common elements in the Universe, so planets that formed around most stars in our galaxy will have plenty of $CO_2$. Volcanoes will keep replenishing their atmosphere, even without any plate tectonics. In our own Solar System, Venus and Mars have atmospheres dominated by $CO_2$. Our Earth would too, if not for the limestone and oceans that keep much of the $CO_2$ locked up. And although life (most in the form of seashells) takes an active part in the $CO_2$ cycle today, the cycle would go on happily without it. What the $CO_2$ cycle thermostat needs is liquid water and tectonic plate activity. Mars and Venus seem too small to have kept their water from escaping and the tectonic activity going. Our Earth barely did! Super-Earths would have an easier time keeping both, and therefore provide long-term stable environments.

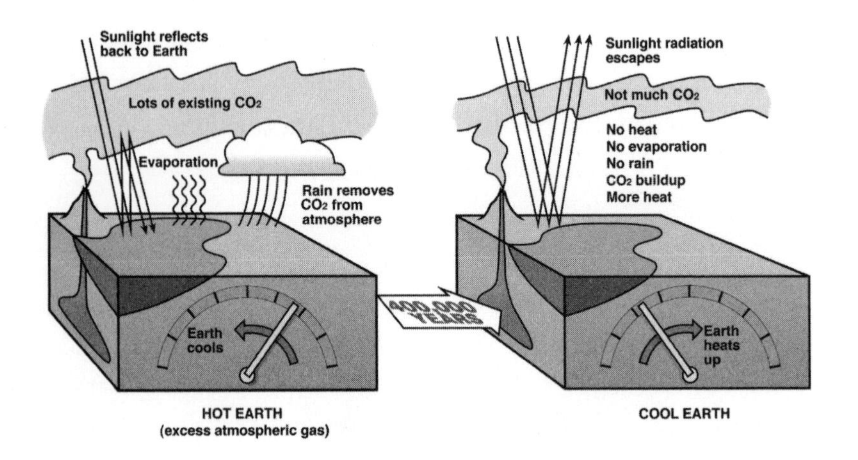


**FIGURE 10.3.** The $CO_2$ cycle functions like a thermostat for the climate on Earth and on Earth-like planets. If the temperature gets too hot (left panel), more greenhouse $CO_2$ gas is removed and the temperature drops; if it drops too low (right panel), $CO_2$ accumulates and warms up the planet.

Corroborating the view that Earth is a planet that just barely supports active plate tectonics is research that shows long periods of slowdown or downright stagnation during its geological history. Certain elements have been depleted from Earth's mantle, which can be taken as a proxy for plate tectonic activity.[10] As the Earth ages and cools, its interior gradually loses certain elements through the cycles of outgasing and subduction. Silver and Behn compared ratios of niobium to thorium and of the two isotopes* of helium ($^4He/^3He$) in Earth's interior, and found a cyclic (instead of a gradual) change since the formation of the Earth. Their interpretation is that plate tectonic activity on our planet slows down and even ceases occasionally, then picks up again. The cycle seems to have occurred at least a couple of times in the past 3 to 4 billion years.

In my work with Diana Valencia and Rick O'Connell we investigated whether plate tectonic activity on super-Earths would be higher, lower, or the same as that on Earth. The answer was not obvious—nothing seems to be simple when it comes to plate tectonics. On the one hand, we reasoned that a super-Earth is hotter inside (over many billions of years) because it is bigger. Higher temperatures inside mean more "boiling," more motion and energy in the mantle, which would lead to more pressure, pushing, and stress applied to the crust from below. Naturally, that leads to breakage as the pieces are pushed around, up and down, leading to the subduction of

---

*An isotope occurs when the atomic nucleus of the same element (same number of protons) has a different number of neutrons.

heavier ones under lighter and less dense ones. Unfortunately, there is another effect we need to consider. The viscosity of the flowing mantle depends strongly on the temperature—just as honey, when you heat it, flows and slips more easily. The problem is this: though super-Earths are hotter and the mantle boils faster, the hotter mantle is also less viscous and slips by the solid crust more easily, without pushing it along. As scientists would say, the two effects seem to cancel each other out. When we looked more carefully at what was going on, we found a third effect that made all the difference. The higher temperature was keeping the super-Earths from growing a thick crust. Finally we had a very robust result for all super-Earths—decreasing crust thickness and increasing mantle flow pressure collaborate to produce a vigorous and "healthy" plate tectonic activity.[11]

Comparing Earth to the theoretical models of super-Earths of different sizes, we find a rich diversity of stable Earth-like planetary conditions. In fact, we find a family of planets that barely includes the Earth, which just qualifies by means of mass, tectonic activity, or long-term temperature stability. Being smaller, Earth is more vulnerable to any number of cosmic accidents. Our biases for our home planet notwithstanding, in this family, Earth is certainly not the preferred child! This is bad for Earthly chauvinism, I suppose, but it's great news for life in the Universe—there seem to be plentiful good places for it.

The question before us now is, What is the current planet census? How big could the family of life be?

## CHAPTER ELEVEN

# THE PASSAGE OF TIME

## *The Universe Is Young, Life Is Younger*

In our story, understanding the passage of time is essential to understanding life in the context of the cosmos.

Next to the building in which I work stands the Harvard College Observatory, built in 1839. It sits atop a hill above Harvard Yard, although you can barely notice the hill. Tall trees and buildings have grown up all around. Much has changed over the years due to population growth, building, landscaping, and so on. And yet something, it seems, has not: the ancient granite rocks of the hill.

Except we know that they have. Tectonic activity, which we on this planet are lucky to have, has actually moved the rocks west by 3.2 meters (the length of my car) since the observatory was built, along with the rest of Massachusetts and the entire North American landmass.

At that rate, the continents would be rearranged completely in about 200 million years. Furthermore, we can confirm this movement with independent evidence from the study of layers of rocks around the world. It paints an ever changing geographical map of planet Earth with cycles of continental rearrangements.

When I am faced with a seemingly unchanging fact—for example, the geology of the Earth's mountains—I try to transform myself mentally into a being that lives to be a billion years old. One heartbeat is now 1,000 years, and 80,000 years would seem like a minute. I am hovering above the Earth among the weather satellites in a geostationary orbit. What does the Earth look like to me? As I watch, the continents are completely rearranged—large ones breaking apart, flowing away from each other, colliding gently with each other, merging again; mountains forming, folding, and being eroded in the process. Through my new eyes the home planet is no longer solid and unchanging but as flowing and dynamic as a pot boiling on a stove.

Such a perspective on the passage of time is less fantastic than it might seem to you. Just think of the moth that lives for one day. In its eyes the green meadow and the lush forest must be eternal. When it comes to the cycles of planet Earth, we humans, like that moth, can perceive the wind rustling in the leaves but not the passing of the seasons. Life on Earth has existed for about 4 billion years and has followed the geological "seasons" and planetary transformations.

If we think about the history of life, not on our timescale but on the timescale of life itself, important facts become obvious. For example, life has existed and developed on Earth for a time—about 4 billion years—that is comparable to the age of the Universe—about 14 billion years. This is a *very significant* fact, what scientists call a nontrivial fact. It tells us that the emergence and development of life is a process on a par with processes of formation and development of planets, stars, and even galaxies. In a certain sense, it makes life appear more like a "normal" cosmic process.

In contrast, we could not say the same about humankind because the timescales are not similar. The oldest known human remains (genus *Homo*) are about 2 million years old. What makes humans different—the appearance of language and technology—is much more recent, both having probably emerged just 40,000 years ago.

The disparity between those two timescales—the human and the cosmic—is great. The ratio between the history of life on Earth and the history of modern humans, for example, is 100,000:1. This is a *very significant* fact as well. It can tell us one of two things: (1) Either the very brief process (development of human society) that emerged from the very long process (development of Earth life) is not comparable to the process of planet development; or (2) we are very lucky to be at the very beginning of a new but short-lived process. The latter case also means that we have little predictive ability.

Predictive ability is important in science because in most cases it means a solid understanding of what is going on. Because we understand the laws of gravity, we are able to predict the future position of the Moon in its orbit and then launch a rocket and land on it precisely as planned. Option 2 is an unfortunate caveat—comparing timescales clearly has its limits.

On the other hand, option 1 is tantalizing! It tells us that a planetary process (life itself) was necessary to develop a life-form (humans) capable of transcending the planetary time-scale. It tells us that life might be a cosmic phenomenon that develops on planets over planetary timescales but leads to forms that are no longer coevolving with the planets—that become independent of the timescale and create their own, along which they evolve, or at least change their environments. However, it takes the process a long time to build up to that level. Option 2 suggests that humanlike life is part of the planetary timescale and consequently represents its final phase but may not last long enough to fulfill its cosmic potential.

At least the potential does exist—humans demonstrated it by going as far as the Moon. A colleague of mine at the California Institute of Technology likes to say that if a process is allowed by the laws of physics, then somewhere in the Universe that process is happening. All kidding aside, there is a deep truth in how we judge potential and plausibility. To me, the fact of our existence on planet Earth today, even with merely the potential to transcend the planet's existence, already implies that this *can* happen somewhere (and sometimes!) in the Universe, even if *we* fail to survive. Whether that transcendence can happen depends on whether our Universe has the potential for future life. Are life's origins at their peak rate, or in their decline? Or, perhaps, just getting started?

Scientists do not know how to answer this question yet, but now they know something about the development of environments hospitable to life. Considering life as a planetary process enables them to make new predictions about its future.

The hierarchy of timescales that involve life is very interesting, and they intertwine with the spatial scale of life. The large molecule scale, as we saw, is on average $10^{-7}$ meters, and the chemical reaction of replicating one unit of DNA lasts about $10^{-3}$ seconds.[1] This is slow compared to what can be accomplished by atoms in the tiny volume defined by $10^{-7}$ meters. However, it is extremely fast compared to any planetary process. This means that life as a process (or sum of processes) will have time to adapt to, coopt, or simply survive whatever it is that happens on the longer, geological timescales, such as the ups and downs of global temperature or the rearrangement of continents. But a chemical reaction is a chemical reaction, and a planetary geochemical process is most often simply the sum of the myriad chemical reactions underlying it. So all those individual geochemical reactions would appear to have similar (short) timescales to the individual biochemical ones that make up the processes of life.

How, then, could processes of life rise above the destructive chemistry of the planetary environment? If it takes the same time to do them, then ordinary chemistry wins out. For life and its biochemistry to prevail, their timescales should be shorter. Life on Earth is competitive that way by using two tricks. One is getting help from special molecules (catalysts, typically called enzymes) to speed up reactions; the other is by keeping tabs on how it does that. In other words, biochemistry has neatly ordered sequences of reactions that do something very well and fast (like storing and releasing energy, forming an enclosure, etc.). In addition, a special molecule keeps tabs on the ordered sequences so they don't have to be reinvented each time. We know such molecules—we all have them and we call them DNA and RNA.

The genetic molecules of Earth life, DNA and RNA, are unique and common to all of Earth biochemistry. Their complexity is the result of a long process of evolution. Interestingly, a molecule can have a structure that encodes a sequence, and that code can be copied and inherited. Equally interesting is that RNA—as has been shown on multiple occasions in laboratories—is capable of catalyzing its own replication. The result is a dramatic shortening of the biochemical timescales—fast rates that can rise above the destructive chemistry of the planetary environment.

Most objects in the Universe that retain their identity over long periods of time are either very big (such as galaxies) or very stable (such as stars and planets). Our Sun will be a star to a venerable age of 10 billion years by being very thrifty in how it spends its energy; it is very stable indeed. But life, thanks to these two tricks it has, presents a third way. It endures thanks to its individual, short-lived, and localized units—organisms that have the flexibility to adapt by doing chemistry faster than the changing environment—that are nonetheless balanced by longer-lived and global entities such as entire populations or species. These larger groups are flexible and allow various members to try different things to survive. That is an extraordinarily smart invention! For all we know, thanks to this, life may be a cosmic phenomenon that, once it has emerged, can continue for an indefinite time.

Humans live short lives, but as a species we have always thought and planned for the distant future. In the past, this might have meant simply caring for offspring who would outlive us; increasingly, we plan for the future as a society. This capacity—underlined by our ability for abstract thought that can reach beyond the horizons of space or time—is perhaps our most remarkable trait. Microbial life may be able to survive most of the slings and arrows the Universe can throw at it, but as we've seen, the Sun will someday put an end to life on this planet. If anything will enable life to endure past the limited lifetime of the planets, it will have to be our ability to think.

There are even bigger implications to the argument that life is a planetary process. We often imagine our place in the Universe in the same way we experience our lives and the places we inhabit. Just as it is easy to think of the rocks at the Harvard College Observatory as static objects, we imagine a practically static eternal Universe where we, and life in general, are born, grow up, and mature; we are merely one of numerous generations, but the Universe itself is still immeasurably older.

This is so untrue! We now know that the Universe is close to 14 billion years old and that life on Earth is 4 billion years old: life and the Universe are almost peers. To put it in more human terms, if the Universe were a fifty-five-year-old, life would be a sixteen-year-old. What's more, the Universe is nothing like static or unchanging.

All of this brings us back to the question, What is our place in this young world? This is a profound question, and there are many ways it can be asked. One of them is simply, Are we alone? I am going to touch upon that question in just one aspect. It has something to do with the recent realization that the Universe is young and is still actively undergoing changes.

The answer to the question could be yes, for a number of different reasons. For one, we (life, not just humans) may be alone because life is an exceedingly rare event, and in 13.7 billion years of history of the Universe we are *it*. On the other hand, we may be alone because we are latecomers to the party. After all, almost 9 billion years passed before our Sun and Earth formed, and so life could have already emerged and died out elsewhere in the Universe, without our knowing it. Or we may be first!

Central to this discussion is the so-called Fermi paradox, named for the renowned physicist Enrico Fermi, who asked the question, "Where are they?" Beneath this question lies the assumption that if there are advanced civilizations out there, astronomers ought to notice them, because surely any advanced civilization would have the power to alter the galaxy sufficiently for us to see. Fermi argued that given the old age of the Universe and the short timescale it took humans to develop technology, other origins of life and civilizations in our galaxy that had a head start should be significantly more advanced than we are. Being significantly more advanced, they would need huge energy resources on the scale of stellar systems and galaxies, which we couldn't help but notice. If we have not noticed anything yet, then, it follows that we may be alone in our Galaxy and technological civilizations must be a very rare occurrence. (I am reminded of Arthur Clarke's statement: "Any sufficiently advanced technology is indistinguishable from magic," which makes me less confident that we know what to look for.)

In the 1990s Paul Horowitz of the Harvard physics department recorded the recollections of Herb York and Phil Morrison about the origin of Fermi's famous question.[2] It was the summer of 1950 at Los Alamos, where a number of American physicists had reassembled, a few years after the Manhattan Project, to develop the hydrogen bomb. Fermi liked to ask rhetorical questions during the group's lunches and then proceed to answer them. So at one of these occasions, according to York's recollection, he asked his table mates, "Don't you ever wonder where everybody is?" Fermi argued that given the large number of stars and planetary systems in the Galaxy and their relatively old age, if life arose and acquired technology elsewhere, the others would be far more advanced and would have colonized the Galaxy by now.

Fermi's conclusion is very sound statistically, as Michael Hart showed in the 1970s.[3] However, the statistical argument is strong only if the timescale of emergence of complex life is much shorter than the age of the Universe, and not so if the two are comparable.

Fermi made his point in 1950, and Hart in the 1970s. In both those eras, the consensus of my fellow astronomers was that the Universe was much older than 10 billion to 15 billion years. The estimate then was more like 20 billion to 25 billion years, and some even argued for a steady state, eternal Universe. At the same time, the geological timescales were already well established at about 4 billion years.

A lot has been learned since then due to an unprecedented revolution in astrophysics at the end of the twentieth century. What scientists have established in the past ten years can help us address Fermi's paradox and the future of life in the Universe. A lot of the history has been pieced together nicely, and for most events we have direct evidence. The story goes as follows:

Light traveling at its limited speed is a great time machine; astronomers train their telescopes on very distant objects and get to see them as they appeared in the historical past. So, when we look back into the sky's past, we see a time about 13.7 billion years ago when the entire observable Universe was made of hot hydrogen-helium gas with tiny trace amounts of lithium. None of the familiar objects of our night skies—galaxies, stars, planets—existed. More importantly, neither did any other chemical elements.

Astronomers can observe that era directly, using a sensitive heat-measuring device that allows them to observe the cosmic microwave background radiation, or CMB. The CMB is the relic light released when the Universe's entire inventory of hydrogen was formed, as previously superenergetic particles combined in atoms. It took merely 20,000 years for this to happen and the light to be released. That light—most of it—has been traveling through our expanding and cooling Universe ever since. Today it is diluted and shifted to longer wavelengths, so what used to be visible light became microwaves and radio waves.[4]

The CMB carries a treasure trove of information via its temperature, temperature variations, and polarization—a subtle measure of how the CMB waves are twisted. These measurements are very challenging and only in the past decade has technology progressed enough to allow such studies, both from the ground and with space missions like COBE, WMAP, and Planck.[5] These direct measurements show clearly that 13.7 billion years ago the Universe had no building materials for life or even for planets—just hot hydrogen and helium gas.

Before continuing with the rest of the story, I need to address the timing of different events in the early history of the Universe. The age of the Universe is known as approximately 13.7 ± 0.1 billion years, meaning that our measurements can't tell if the age is 13.6 billion or 13.8 billion years, or anywhere in between. At the same time, some events can be timed more precisely in relative terms. Therefore it has been easier to refer to the times for different events, as times *since* time-zero (called the big bang). For example, the CMB was released 379,000 years after the big bang. This timing of the CMB is a measurement and the preceding statement remains true regardless of whether the age of the Universe (i.e., the time of the big bang) is 13.6 billion, 13.7 billion, or 13.8 billion years ago. Alternatively, if we fix the big bang at 13.7 billion years ago, this same event (the creation of what has become the CMB) occurred 13.6997 billion years ago.

Now, back to our story. The obvious question we have to answer is, Where did the building materials—all the chemical elements like carbon, oxygen, silicon, and iron—come from? The answer is well-known—they all came later, from stars. That brings us to the next notable event in our Universe—the formation of the first stars.[6] This is an event that we do not yet see directly, although the successor to the Hubble space telescope is being built to do that. Nevertheless, scientists already have plenty of indirect evidence that this happened about 13.1 billion years ago.

Stars, including the first stars, are very unusual objects when you look at the big picture. They are stable and long-lived concentrations of ordinary, or baryonic, matter. There is nothing unusual about that. Ordinary matter is found all over the Universe (billions of galaxies' worth) in big and small clumps that just sit there and do nothing—except when some of these clumps get compressed under their own weight and form stars. It just happens that the balance between gravity pull and matter repulsion is achieved at temperatures and densities inside the star that allow the atomic nuclei of hydrogen and helium to fuse. When you fuse atomic nuclei, two important consequences follow: lots of energy is released and new, heavier nuclei are formed. That is how our Sun shines.

Stars are the queens of fusion—they do it admirably well! They literally light up the place and proceed to transform it from a boring simple gas to the richness of the entire table of the elements.[7] The process is orderly: first, hydrogen fuses into helium until the central regions of the stars are chock-full of helium, which, being heavier than hydrogen, shrinks and heats up. Helium heats up until its threshold for fusion is reached, and then a new stage in the life of the star begins, at least inwardly.

While fusing hydrogen produces mostly helium (fusion would be a clean, powerful source of energy for humankind, if we ever learn to do it in a controlled fashion), the fusion of helium produces a number of heavy elements, most notably carbon and oxygen. Stars can fuse elements all the way up to iron, at which point they stop, lacking sufficient energy to go any farther—unless the star is big enough to explode. In such a supernova, more fusion can happen that produces many more elements and frees the rest to capture electrons and become the atoms of heavy elements with all the rich chemistry they can cook up.

Astronomers can observe how the stars enriched the Universe in heavy elements. The large telescopes of the past ten to twenty years have allowed them to peer back to about 12 billion years ago. They see some heavy elements, such as iron; they see patterns in which elements are relatively enriched and that reveal how stars produced them. The picture that emerges is one of generations of stars steadily transforming the hydrogen and helium of the young Universe into all the heavy elements.

Stars form out of low-density gas, which must cool while being compressed (under its own weight) for a star to be able to condense. Because hydrogen and helium are terribly bad at such cooling, the first stars must have been super-size only—hundreds of times larger than our Sun. The first stars were massive, had short lives, and produced some heavy elements that were dispersed inside the nascent galaxies and made it possible to form smaller, less massive stars. The addition of even a sprinkling of elements to the hydrogen-helium gas helps it cool, so the next generation of stars can be formed from a wider range of gas clumps. With each generation, progressively smaller stars can form—and they do. Today our Galaxy has many stars smaller than our Sun. Partly this is due to the fact that smaller stars live longer by burning their nuclear fuel slowly, but mostly because small stars have formed in increasingly larger numbers as the Universe has evolved.

Small stars disperse a portfolio of heavy elements when they die, so the enrichment of the Universe with heavy elements continues at a slow, steady pace. In fact, after 13 billion years only about 2 percent of the original mixture has been transformed to heavy elements; the enrichment, as astronomers like to call it, is very slow indeed.

The brief story of the Universe, then, looks like this: from just hydrogen and helium about 13 billion years ago, generations of stars made enough iron and oxygen, silicon and carbon, and all other elements, to be able to form Earths and super-Earth planets. There are at least two important morals to this story regarding life.

First, it took a long time before stars anywhere in the Universe could have planets. Stable environments in normal galaxies that were enriched enough to have planets became available about 9 billion years ago.[8] If you ask about large terrestrial planets, such as rocky super-Earths and Earths, then it is more like 7 billion to 8 billion years ago. We can imagine that the emergence of life had to wait until that time in the history of the Universe, if not later.

Second, the enrichment continues to this day, and we have a fairly clear idea of how our Universe will be transformed in the eons to come. For example, we see that massive stars have been forming less frequently for the past 5 billion years, so the small stars will dominate element production and enrichment in the future. Generally, that means more carbon than oxygen. Today there are three times more oxygen atoms than carbon atoms in most of our Galaxy, but eventually a point will be reached when carbon and oxygen exist in equal abundance. When this happens, the mineralogy of rocky planets changes. Carbides dominate silicates, and there will be important implications for the origins of life on such planets, as the carbon planets described earlier in the book go from being rare to being common.

In general, though, the future of life looks excellent. Unless life is an exceedingly rare phenomenon, there should be more of it, and more diverse forms of it, in the future. Planets may be just a tiny fraction of the Universe because they are so small, yet there are so many of them that there are plenty of places for life. We now know that our Universe is passing through its peak of forming stars (known as the stelliferous era), but it appears that it is still peaking in terms of forming planets.[9]

This implies that the Fermi paradox, which is about the past, is the wrong way to look at the question of whether there is life elsewhere. The paradox assumes that there was enough time before us for others to emerge and develop. The new evidence does not support such an assumption easily. Of course, when it comes to technology, not microbial life, we can only speculate–our own technological capabilities have grown exponentially recently, and if such growth were used as a basis, then the Fermi paradox remains strong statistically. But for life, the logical sequence I follow is: (1) complex chemistry is necessary for life to emerge—enough heavy elements are needed; (2) stable environments that allow chemical concentration are also necessary—terrestrial planets (Earths, super-Earths) are needed. When in its past did our Galaxy (and our Universe) fulfill these requirements?

The answer is, Between 7 and 9 billion years ago. I arrive at this answer via two independent paths. The first path, much of which relies on what we've been considering, is to observe the stars and gas in distant galaxies, measure their abundance in heavy elements (the ones needed for life and planets), and thus see how their abundance grows with time. When we begin seeing stars with just enough heavy elements to allow forming Earths and super-Earths, we have pinpointed the time in the past we are looking for. The only problem is that we need to know how much heavy elements are enough to form big terrestrial planets. That's a tough question. If our computer models for planet formation are accurate, then a solar system requires at least 1/1,000 of the proportion of heavy elements that our Sun has. Our Galaxy reached this state about 9 billion years ago.[10]

The second path to answering the above question goes directly to the planets. Do we observe a decline in the number of planets around stars that are poor in heavy elements? Yes. This evidence surfaced early on in the planet-hunting game. It was so pronounced that most teams were tempted to select stars rich in heavy elements in order to discover more planets. Nobody was surprised that such a trend—more metals, more planets—existed, but the strength of the trend was surprising. The trend drops off to practically no planets so fast that even the proportion I mentioned above—1/1,000 of the heavy elements of the Sun—seems too generous. It comes out to something like 1/100 of heavy elements compared to the Sun.[11] This would put the time in the past when planets that were capable of cradling life could form at just about 7 billion to 8 billion years ago.

A word of caution is due here. The "more metals, more planets" trend is currently only observed for Jupiter-like and Saturn-like planets, and for hot Jupiters in particular. I have to assume for now that it holds for terrestrial planets, but the Kepler mission is working to answer that question accurately.

Today astronomers know with certainty that less than 13 billion years have passed since our Universe was capable of having stars and planets. This makes the stellar, planetary

Universe very young. (Because we see that our Galaxy and the rest of the observable Universe, and its 200 billion galaxies, show a clear potential to continue on as we see them today for hundreds of billions of years, if not much longer, I feel that the words "very young" describe the Universe adequately.) The anthropomorphic analogy to parent-daughter, when we talk about a Universe with planets and Earth life, is then pretty good, as well. Life on Earth could really be among the first older siblings in the family.

So far I have been talking about microbial life. But what about the bigger question: Are we humans alone? That is a far more difficult question to answer. However, if planets and life are so young in our Universe, perhaps we are not latecomers to the party. We may be among the early ones. That could explain why we see no evidence of "them." This does not necessarily mean, however, that no one is there.

By all accounts, today the Fermi paradox remains unresolved and allows for a fascinating range of possible solutions—from the very deep to the very entertaining, all of them worth more attention than I plan to give them here, but for recommending the rich literature that does.[12]

With this answer to the Fermi paradox in hand, we can now estimate just how big the family of life—the census of habitable planets—is. The answer is, Pretty big. Consider this: there are more stars in the Universe than there are grains of sand in all the beaches on Earth.[13] And there are equally as many planets (see Figure 11.1). Of course, as I noted at the beginning of this book, those astronomical numbers do not imply inevitability, no matter how good we feel about our models. We have to go

Galaxies in the Hubble Ultra Deep Field
Hubble Space Telescope
NASA, ESA, and R. Thompson (University of Arizona)

Our Milky Way Galaxy

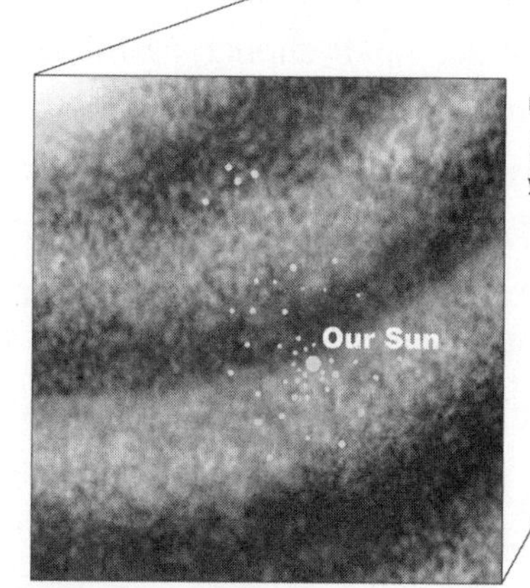


1,000 light-years

**FIGURE 11.1.** There are as many grains of sand in all the beaches on Earth as there are stars and planets in the Universe. The bright dots in the vicinity of our Sun denote some planetary systems we have already discovered.

and find out for ourselves. The survey by the NASA Kepler mission will accomplish that. In the meantime, we can use the current discoveries of extrasolar planets to make a preliminary estimate.

To begin, the number of known planets to date, summer of 2011, is in the mid-hundreds (about 600), and most of them are in our neighborhood of the Galaxy (see Figure 11.2). They will be a useful reference.

First, I need to know the number of stars in the Galaxy. This number is being constantly updated but has not changed much in the past decade, and is based on many different surveys. Many millions of stars, of different types and in different parts of our Galaxy, have now been counted. With these counts and a measure of the extent of the Galaxy, I multiply to obtain the total number of stars: about 200 billion stars in total. Of these only 90 percent are small enough and long-lived enough to develop and have planets. In addition, only 10 percent of these smaller stars were formed with enough heavy elements to have Earth-like planets. So far, our estimates are very secure and robust. But now I need to know how many of that 10 percent of stars actually harbor Earth-like planets.

I turn to planet counts, just as with the stars. I count how often Earths and super-Earths pop out in the planet surveys done so far. This is a difficult task because few super-Earths have been discovered (no Earths to date); as we've seen, they are much more difficult to find, compared to the large and heavy giant planets, so any census of them needs to take this difficulty into account. One way to do that is to compare two different methods of planet discovery and see if they lead

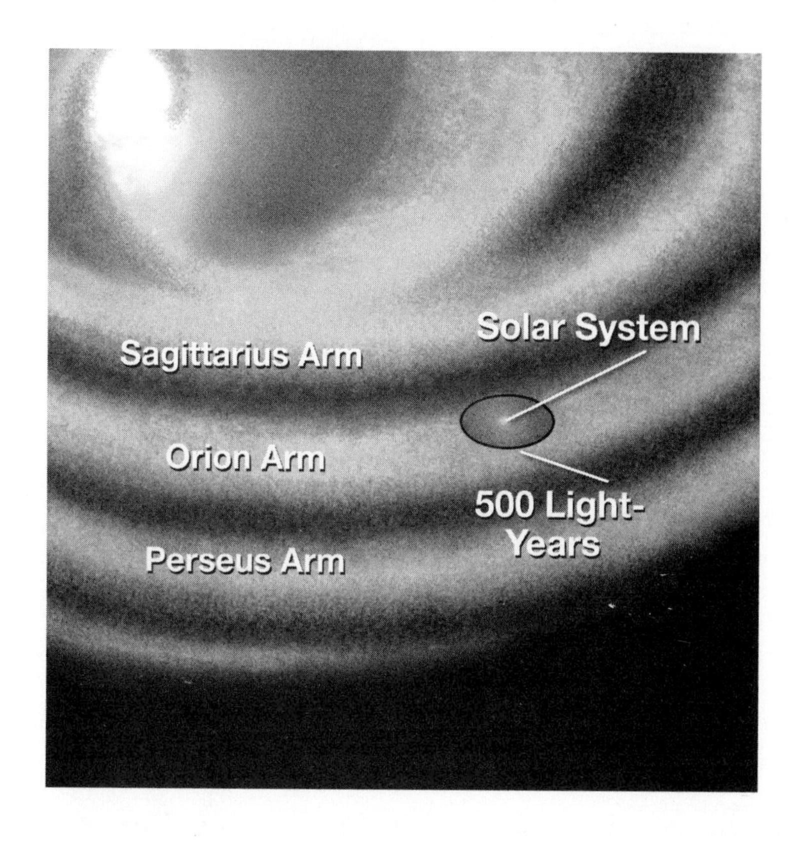


**FIGURE 11.2.** Most extrasolar planets discovered so far are close to our Solar System; about 80 percent of all known planets are within 500 light-years and are around stars from the Orion arm of the Milky Way.

to the discovery of different ratios of super-Earths to giants. Comparing the Doppler shift method with the gravitational lensing method suggests that about half of the stars with heavy elements should have at least one Earth or super-Earth planet. If we assume that there are no privileged orbits—that all else being equal, planets can form and remain in any orbit around a star—then only about 2 percent of these Earths and super-Earths will happen to be inside their star's habitable zone. The remaining 98 percent will orbit too close or, more often, too far from their star. The ultimate answer will come from the Kepler mission, near its nominal end in a few years, but the preliminary data already is consistent with this estimate.[14]

Now I am ready to sum up the numbers. With all these fractional reductions to the population of 200 billion stars in our Galaxy, I end up with 100 million planets with habitable potential today.

This number is not precise, but there is no escaping how big it is. On the other hand, the majority of these planets are of an age similar to our Earth's. Some are younger, but only a few, given the recent advent of heavy elements, should be much older. With such time constraints, we need to be cautious about drawing conclusions of inevitability. All we can be certain of is that life is not an impossibly rare phenomenon—it definitely has odds of 1 in 100 million. These odds aren't so bad; events of such rarity do happen. As of May 2011, the US Megamillions lottery jackpot had been won four times that year, with odds of 1 in 176 million. In fact, using the binomial distribution, we can see that there is an 18 percent chance of two successes in 100 million trials.

As I explained in the previous chapter, we live in a very young and *changing* Universe, so the estimate of 100 million planets with habitable potential is just a snapshot of our Galaxy now. A real estate developer once asked me about the trend in the number of habitable planets—are they growing or diminishing in number? Good question!

The stars tell a story. About ten years ago, thanks to the Hubble space telescope, it became possible to measure many very distant galaxies, looking back through time in the process. Astronomers could look at the colors of those galaxies. The color of a galaxy is an aggregate of the colors of its myriads of stars. When this was done carefully, with astronomers making sure to note "redness" due to dust, and so on, a fascinating pattern emerged.[15] The colors of galaxies got "bluer" early on in their existence, then peaked about 7 billion years ago, when they began getting redder very quickly.

What do these changing colors mean? Galaxy colors reflect the colors of their constituent stars, so we have to talk about star colors. For about 90 percent or more of its life, a star's color tells us how heavy that star is. As we saw in the last chapter, the more massive a star is, the faster it fuses hydrogen. That makes such stars hotter and more luminous. The extra energy they possess causes them to shine with bluer light, just as the filaments of an incandescent bulb do relative to an electric stove heater. Burning off faster also means that bigger stars have short lives.

So, then, if we see "bluish" galaxies, we can conclude that they have plenty of big, heavy stars and that such big stars are being formed at a high rate. If not for the latter, "bluish" galaxies would be very rare. Therefore, the colors of distant galaxies can tell us the rate of star formation.

In the Hubble images, the changes of galaxy colors with time show us the changing rate of star formation. Stars and galaxies formed early on, and then stars kept forming at a steadily increasing rate during the following 5 billion years. In the past few billion years that rate has plummeted, worse than the 2008 world markets downturn and with no prospect for recovery. The heady days of big stars are over! Our Galaxy and other galaxies in today's Universe form ten to fifty times fewer stars, and most of the stars are small.

This history is good for chemistry and for life. The bumper crop of early, short-lived stars enriched the Universe and the next generation of stars (and planets) with heavy elements. A galaxy with fewer big blue stars is better for chemistry and life too. X-rays and UV light, which large, "blue" stars create in large quantities, are health hazards. In the past 7–8 billion years our Universe has become more and more hospitable, and will continue to evolve in this direction for a long time in the future.

In the meantime, there are at least two reasons why we should be optimistic about the plurality of worlds with separate origins of life in our Galaxy. First is the apparently short time it took life to emerge and take hold on the surface of the young Earth. Second, we now know that the family of hospitable planets is even larger than we thought, with super-Earths being excellent cradles for life.

## CHAPTER TWELVE

# THE FUTURE OF LIFE

Earth is our home planet. We call it the cradle of life. These notions are deeply ingrained—but wrong. A home is a place we live in, grow up in, and leave; we buy and sell them. A planet is not a home to life. Rather, the planet and the life on it are the same thing. If I were to leave this discussion here, an astute reader might accuse me of making a trivial point, akin to the politician who refers to the same thing by a different name in order to tax it more. After all, life is a planetary phenomenon, albeit not as transient as a hurricane or as inconsequential as a continental tectonic plate. Yet even the single example of life we know has the property of being transplantable (to another "lifeless" planet). Or at least we think so.

How can this be so? As I suggested in my parable of the three planets, we already know that the chemical process known as life is sufficiently powerful to transport itself from Earth to the Moon, and there's a good chance that we've already sent bacteria as passengers on our spacecraft to live permanently on Mars—and maybe far beyond, whether on Titan thanks to Huygens or via Voyager to who knows where. But this kind of transplantation of microbial spores could have happened long before we began exploring space, and over greater distances. When Earth was being bombarded by asteroids early in its history, the outflow of material would have been nearly as great as the inflow, and it's not hard to imagine some of that material being life, ultimately to travel interstellar distances, embedded in dust particles or comets, until it found a new home on some other habitable planet. In that way, we can imagine planet Earth as a home or a cradle for life in the conventional sense. The idea is old and is called panspermia.

Panspermia, which means "seeds everywhere," is an old idea that can be traced back to the Greece of Socrates. Svante Arrhenius, a Swedish chemist, revived the idea in the early twentieth century. Additional arguments were put forward by Fred Hoyle and Chandra Wickramasinghe in the 1970s, with comets as the delivery vehicle for spores, in order to protect them from radiation damage. (Hoyle and Wickramasinghe were also proponents of an eternal steady state Universe, with plenty of time for any spores to make their journey across our Galaxy and mix well among its stars. That kind of thinking made the Fermi paradox seem very appealing.) There is also a local version that postulates exchange within a planetary system, such as from Mars to Earth. The idea goes as follows: a large asteroid impacts young, wet Mars and ejects debris into space and some into independent orbits around the Sun. Though such impacts resemble dramatic explosions (they melt and vaporize huge amounts of rock), many of the ejected rocks should remain entirely unheated and undisturbed. If Mars was alive, with microbes deep inside rocks, many microbial colonies should have survived inside large chunks of rock.

With time—millions of years—the orbits of small rocks are perturbed by the changing gravitational pull of the planets. Often those orbits change dramatically and cross the orbits of large planets. Some of them end up on the surface of a neighbor planet as meteorites. There are many Martian meteorites on Earth; they are relatively easy to identify by their minerals and inclusions of small bubbles containing Martian atmosphere. This suggests that if life ever existed on Mars, it could have reached this planet too. Interestingly, it is far less likely to have gone the other way, as the predominant traffic flow is from the outer Solar System in. When orbits change, they tend to preferentially move toward the Sun.

Whether this has happened is simply an empirical question because the mechanism itself is quite plausible. We know that there are microbes durable enough to make the trip into space and endure the low temperatures and high radiation, even over the million years required for a typical "meteoritic" trip from Mars to Earth. Similar trips between stars, though possible, would take much longer.[1] Even if spores might survive billion-year journeys (no such spores are known today), the question remains if there has been enough time in the history of our Galaxy for such journeys to be completed yet. As noted in previous chapters, our Universe is young and the components needed for life are even younger. Just as there has been barely enough time for life to have evolved at all, there is even less for spores to have dispersed through the volume of our big Galaxy. Even if panspermia is not the state of life today, it surely could be in its future.

Panspermia is a long-term phenomenon, however, and so the future of life, at least on our planet, may seem doomed to go on much as it has in the sense that evolution will continue operating on the same basic rules, with the same basic tools, such as DNA and RNA. But that scenario, I think, is far from the truth. In fact, we are living through, in the first years of the twenty-first century, a remarkable turning point in the history of life on Earth. For the first time in about 4 billion years a new species is *not* going to emerge from the set of processes that led to the diversity of life on this planet. Instead, one species is going to synthesize another—a life-form that is unique, but not in the way that a new dog breed or a genetically modified corn plant is made unique by some cosmetic differences with its progenitor. It will be new in terms of its unique biochemistry, a new life-form that has no place on Earth's tree of life, a new life-form at the root of a *new tree of life*.

I am describing the dawn of a new field—synthetic biology. All Earth life has a striking unity of shared biochemistry, which is why we can use *E. coli*, fruit flies, and lab mice as proxies for our own biology in the laboratory. Much of the cellular machinery is the same. Through synthetic genomics (or, more generally, bioengineering) scientists use that unity to create diversity in form and function. Think of age-old breeding, or the current promise of fully designed microbial genomes.[2] Synthetic biology, the way I refer to the field here, goes beyond synthetic genomics, both conceptually and in practice. It is no longer a matter of modifying a genome or even writing a new one, but of synthesizing biological systems that do not exist as such in nature and using this approach to understand life processes better.[3] This amounts to changing the basic biochemistry and that is why this new life-form cannot belong on Earth's tree of life.

Defined thus, synthetic biology research is closer to chemistry than to biology, while the opposite is true for genetic engineering.[4] (This led Pier Luigi Luisi to coin the term "chemical synthetic biology" in 2007.)[5] I see as one of its central concepts and objects of study the minimal artificial cell—a hypothetical chemical system enclosed in a vesicle and capable of life's main functions.[6] It is a cell because of the vesicle; it is "minimal" because it is stripped down to a bare minimum of function needed for self-sustained existence and evolution; it ought to be "artificial" because no one expects to discover such a primitive (and vulnerable) chemical system on Earth, where even the simplest microbes are highly sophisticated, complex cells and have occupied every conceivable space and niche. At the opposite end, a genetic engineer works with these same highly sophisticated forms, including plants and animals, and strives to modify them to perform in ways that go beyond their naturally endowed skills.

Synthetic biology brings the prospect that an alternative biochemistry is possible and may have developed independently on other planets (or on Earth). This prospect is remarkable in that it involves ordinary chemistry, as opposed to discovering a new form of matter or a new force. Instead, the anticipated discovery lies in uncovering new basic rules (or "laws") of nature. It is also remarkable because it allows us to ask the origins of life question in a way that may lead science to a breakthrough: What is the role of initial conditions in shaping the unity of Earth biochemistry? Does the diversity of planetary environments map onto a diversity of alternative biochemistries?

This prospect is significant because it pulls together the various threads woven throughout this book, in a synthesis that was hardly possible until very recently. The work in the lab has direct implications for the astronomical search for life on distant planets, and vice versa: these remote findings help hone the lab search for pathways to alternative biochemistries. To find if we are not alone in the Universe, we must understand life, and to understand life, we must learn how to build its chemistry.

When we try to grasp the challenge, the unity of Earth biochemistry can provide a helpful clue. Think of all chemistry—its millions of molecules and all the reactions that can take place among them under different conditions—as a large space. Mathematically speaking, that is a multidimensional space that human brains can't visualize, but if we choose a few relevant parameters and reduce the many dimensions to two or three, we can see patterns that often resemble a geographical map with valleys and mountains. Scientists often call such spaces "landscapes." The chemical landscape is shown in Figure 12.1.

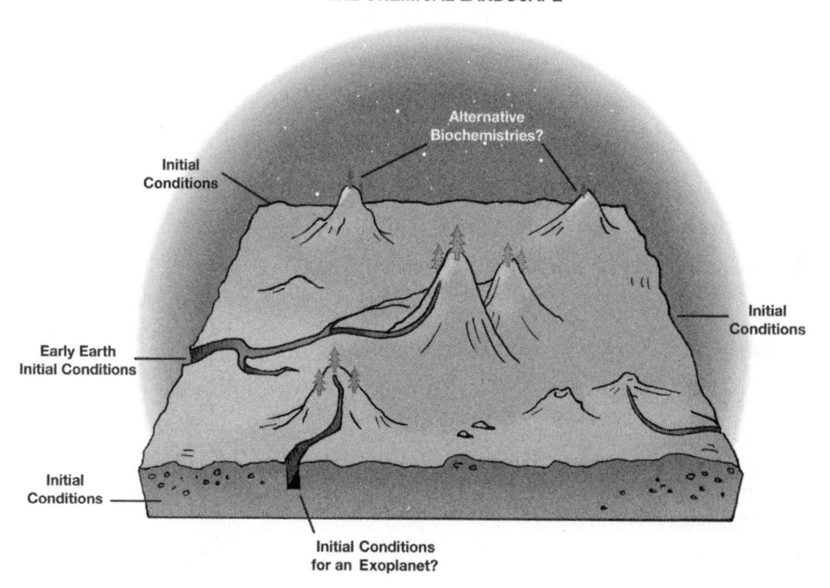


**FIGURE 12.1.** The chemical landscape shown here is a metaphor illustrating the space of all molecules and their chemical reactions, in terms of, for example, molecular reactivity. Surrounding the space are the entry points for assembling molecules (from atoms) under different environments. We call these "initial conditions." Often an initial condition can occur in nature, as on early Earth; often it can only be achieved in a lab. One of the highest peaks in the landscape represents Earth's life biochemistry; a pathway from early Earth initial conditions leads to it. We don't know if other biochemistries exist, nor do we know if multiple pathways reach them or reach Earth's biochemistry (or its mirror handedness version: the second peak nearby).

The most extraordinary feature of the chemical landscape is a high peak—narrow and well-defined—which encompasses all of Earth life biochemistry. The peak is high because life is a highly efficient chemical system today; evolution has raised its height through time. Its narrow base, on the other hand, illustrates how constrained in its chemical choices life is. It uses only twenty amino acids, only left-handed ones, only some proteins, only RNA and DNA, and so on, out of many millions of available choices in the vast chemical landscape. Are there any other tall peaks in the landscape? If so, which of them are alternative biochemistries ("other life")? Are any of them accessible from plausible planetary environments?

We can answer the question about alternative biochemistries in the affirmative because of the twin peak corresponding to "mirror life"—same choice of biomolecules but opposite symmetry (handedness) (see Figure 12.1). There is no reason to suspect that a mirror biochemistry would behave any differently in a lab or on a planet. While scientists understand why a mix-and-match biochemistry fails, the choice of left versus right symmetry might have simply been a matter of chance.[7] In that sense, the mirror biochemistry is trivial—it provides no direct insight whether truly alternative biochemistries are possible. However, if it can be built in the lab, such a biochemical system will deliver powerful tools for understanding life's fundamental principles. For one, it might be the shortcut to the successful assembly and maintenance of a minimal cell that will open numerous opportunities.[8] At least, having a functioning mirror system will allow testing the importance of initial conditions (e.g., different planetary environments).

Chemists today have analytical tools at their disposal that allow them to explore large swaths of the unknown interior territory of the chemical landscape.[9] Similarly, for the first time in history, astronomers have tools to explore planetary environments unknown on Earth or in the Solar System. Soon they will explore and understand alternative global cycles to Earth's carbonate-silicate cycle (see Chapter 10) on Earth-like exoplanets orbiting other stars, and map the hitherto unknown regions of the periphery of the chemical landscape, where the initial conditions reside (see Figure 12.1). Working together, chemists armed with information provided by the planetary astronomers could venture into the interior in search of high peaks (alternative biochemistries). Would a super-Earth planet with a dominant sulfur ($SO_2$) cycle lead them to a new "sulfur-inspired" biochemistry? Or would they find an alternative pathway to our own Earthly biochemistry? Whatever they found, it would be a trip to remember! Perhaps the most valuable thing we'd learn is how to look intelligently for signs of life on those distant planets, as opposed to simply, and naively, looking for carbon copies of Earth's biosphere.

The history of our species, *Homo sapiens*, has several big milestones. We know most of them, perhaps we are missing a few. One thing is for sure: they are all milestones important to *our* history and the history of life on Earth. However, this milestone I am anticipating—synthetic biology—is of a different character and goes beyond the special case of planet Earth and *Homo sapiens*, because it is significant in the chain of events following the development of ordinary matter in the Universe.

Earth life, a single example of the complicated chemistry known as biology, proves feasibility—it is possible for biology to happen, but it proves nothing about how likely it is to happen. Synthetic biology, once its research strategy succeeds, could prove that ordinary matter has an inherent capability to self-organize, to create diversity from a single biochemistry, and ultimately to amplify that diversity by spawning multiple biochemistries. (What we do not know is the magnitude of that amplification.)

That last step is a historic watershed: one tree of life begetting other trees of life (or "roots" of other trees). It is a watershed because it could be a recipe for amplification of diversity on the scale of the Galaxy and on long timescales (billions of years) and because it suggests the existence (now or in the future) of a new generation of life. Let's call it Generation II life. Its defining feature is that its tree is not rooted in prebiotic chemistry but originates in Generation I life. Life on Earth is Generation I, and the term "generation" is used in the same way as in human society. It implies a cohort of peers, constituting a single step in the line of descent from an ancestor, though not necessarily born strictly at the same time. Generation I may consist of a collection of biochemistries, if such exist.

We do not know how different other biochemistries (or other origins of life) could be from our own. We might discover that the "biochemistry landscape" allows only limited and, perhaps, similar families of biomolecules and biochemistries. If so, the amplification factor afforded by synthetic biology will be considerably diminished, but it will still be an amplification factor, and that implies an increasing role of biochemistry in the redistribution of baryonic (ordinary) matter in the distant future of the Universe. Given what we know about what other planets can be like or will be like—as when the carbide planets come to outnumber the silicate ones, such as our own—there is plenty of room in the chemical landscape for the biochemical landscape to be vast. Generation II life may already exist in the Galaxy.[10]

The milestone of synthetic biology is, to my mind, one of three ongoing. They appear unrelated and may have happened at once coincidentally, each a product of the highly accelerated rate of our technological development in the past half century. The other two are the completion of the Copernican revolution and the astonishing process of globalization, as humans across the planet have become interrelated not just biologically, but in practical ways and in everyday awareness.

Globalization has already happened and would have been inconsequential to our discussion here, but for its direct relation to the completion of the Copernican revolution (the former having helped make the latter possible) and its not entirely positive effect on Earth's entire biosphere. I hope that the completion of the Copernican revolution, by showing us that the Earth is just another planet and that other planets may be hospitable to life, will help convince us that we are not special. The humility could do us some good. Looking at life as a planetary phenomenon in which the underlying biochemistry is deeply tied to the planet itself will help reinforce our awareness of being one with our Earth, a product of a unique biochemistry that emerged 4 billion years ago and is distinctly *Earthly*. We are part of a good thing here, and perhaps learning about it will help motivate us not to screw it up.

The dawn of synthetic biology, then, comes at a fortunate time: it answers the question, What next? that emerges after the completion of the Copernican revolution. It transforms the end of a chapter on humankind's awareness of the world into the beginning of a chapter about humankind's place in the world.

In these pages I have painted an optimistic picture in which life is robust and emerges with ease in a Universe full of places where it can grow. We do not really know if life emerges with ease. We only know that it did so here on Earth. One example is not enough to draw conclusions.

I have also suggested that panspermia, whether accidental or purposeful, via rocks, comets, or interplanetary probes, is very possible too. That is one reason why life may be a process that, once it has emerged, can continue indefinitely, never attaining equilibrium with its environment, even on stellar or intergalactic timescales. One piece of evidence for this is us—we know we are life-forms capable of leaving our planet of origin and exploiting other resources. Even if we never leave permanently, the fact that we can do so proves that life is a phenomenon capable of transcending the lifetime of a typical star, such as our Sun.

And a good thing too! Imagine our Solar System 5 billion years from now. The Sun—the parent star, source of light, provider of warmth and energy to living things on our life-transformed planet—is taking a well deserved retirement and is about to begin spending its retirement account faster than a savings-free baby boomer. And the Earth? Well, the Earth will have to go. Venus and Mercury will have to go too—engulfed, molten, and vaporized in the slowly expanding hot sphere of the red giant star that is now our Sun emeritus. Planet Earth and its 9 billion-year-old biosphere are gone for good! The microbes do not have an evacuation plan.

No need to panic, though, since 5 billion years is a bit beyond your retirement age. Nevertheless, the Sun's retirement (like our own) is something we ought to plan for well in advance. And perhaps instinctively, humankind is already on a path to do just that. Understanding the essence of life here on planet Earth will help us understand the origins of life in other places in the Universe. With this knowledge we will seek and find friendly harbors. And one day we will throw anchor there. This is the day humankind, and with it Earthly life, will free itself from the cosmic fate of planet Earth and our Sun.

We humans have made this kind of trip numerous times in our brief history. Here is just one example.[11] About 4,000 years ago tribes in south central Europe had domesticated horses and invented carriages. They could move entire villages over vast distances—a thousand miles, maybe more. After several generations, living conditions in the European steppes had grown unbearable, so one day they packed up and left for the east, where sparsely populated lowlands opened before them. For another century or more—nobody knows exactly how long—these people moved east until they reached the towering mountains of central Asia. Through local tribes they heard of a fertile valley just southeast of the Pamir and Tienshan mountains—hard to reach but uncontested land. These people of the steppes had the knowledge and the technology to cross the high mountain passes, some exceeding 10,000 feet. The other side must have appeared to the worn-out travelers like a place out of this world. Today this is the land-locked Tarim basin in the heart of Asia, mostly a salty sand desert—the Taklimakan. But geological evidence shows that as late as 2,000 years ago the Tarim appears to have been rich in water and vegetation.[12] The tribes from Europe not only survived but prospered. Today we marvel at their exquisite clothing, beautiful artifacts, and rich culture in the amazing mummified burials discovered in the dessicating sands of Tarim in the past decade.[13]

This is just one such story. The time for such migrations on planet Earth has ended. Today the planet is densely populated and globalized. Our planet is a beautiful place and we could be happy living here for thousands of years to come. But we already know that one day our kind will face the same decision the Tarim migrants faced all those years ago. Will our future relatives have the knowledge and technology to make it across?

# NOTES

## CHAPTER ONE

1. For a detailed account of this history, see Charles A. Whitney, *The Discovery of Our Galaxy* (New York: Knopf, 1971).

2. Temperature is measured by different scales—Celsius, Fahrenheit, Kelvin, each with a different zero point. The Kelvin scale begins at "absolute zero," while the Celsius scale has its zero point at the temperature distilled water freezes under sea level pressure. Therefore, 0 degrees C corresponds to 273 K, while 170 K is a very cold minus 103 degrees Celsius. D. Sasselov and M. Lecar, "On the Snow Line in Dusty Protoplanetary Disks," *Astrophysical Journal* 528 (2000): 995.

3. Planets orbiting other stars are named after the star followed by a lowercase letter "b," "c," and so on, in order of discovery. The shortened constellation name (e.g., 51 Peg for 51 Pegasi) is commonly used. Whenever the star has no previous common name, the name of the project responsible for the discovery is

used, followed by a consecutive number and by a lowercase letter "b," "c," and so on.

4. This is the first valid detection of a planet outside our Solar System (D. Latham et al., "The Unseen Companion of HD 114762: A Probable Brown Dwarf," *Nature*, May 4, 1989), but it was not announced as such because the authors of the work were cautious not to overinterpret their evidence. Discovered with the same technique used to find 51 Peg b, the planet's mass is derived only in its minimum limit, meaning that if we happen to be observing the planet's orbit face-on (i.e., from its pole), its mass must be larger. The probability is not negligible, particularly when compounding the case with two unusual properties of the HD 114762 companion: (1) its mass exceeds that of Jupiter, yet its orbit is smaller than that of Mercury, and (2) it has a substantial orbital eccentricity. For comparison, 51 Peg b at least has a noneccentric orbit, though what an orbit it is!

5. G. Walker et al., "A Search for Jupiter-Mass Companions to Nearby Stars," *Icarus* 116 (1995): 359.

6. S. Ida and D. Lin, "Toward a Deterministic Model of Planetary Formation," *Astrophysical Journal* 626 (2005): 1045.

7. The question of the Other has fascinated writers, philosophers, and anthropologists; a nice analysis of Western thought, albeit confined to mostly French sources, is contained in the seminal monograph by Tzvetan Todorov, *Nous et les autres* (Paris: Editions du Seuil, 1989; *On Human Diversity* Eng. trans., Cambridge: Harvard University Press, 1993).

## Chapter Two

1. Dava Sobel, *The Planets* (New York: Penguin, 2005), 145.

2. Helium was discovered remotely in the Sun through spectral analysis of the signatures of gases in solar light, not in a mineral or laboratory on Earth. Hence its name from Helios, the Sun.

## Chapter Three

1. D. C. Black, "Completing the Copernican Revolution: The Search for Other Planetary Systems," *Annual Reviews of Astronomy and Astrophysics* 33 (1995): 359. This fascinating and insightful review was written at the dawn of the age of extrasolar planet discovery. It shows the awesome technical challenges, the frustrations, and the gnawing doubts after the many empty-handed searches. M. Mayor and P.-Y. Frei, in *New Worlds in the Cosmos* (Cambridge: Cambridge University Press, 2003), give an account of the beginnings and provide full quotes from the writings of C. Huygens and B. de Fontenelle.

2. The Hubble space telescope observed the area in the sky known as the Hubble Deep Field for ten consecutive days, taking multiple images in four different filter passbands: near-ultraviolet (300nm), blue-yellow (450nm), red (606nm), and near-infrared (814nm), for a total of 342 individual exposures.

3. There is an extensive literature on direct imaging for planet detection for both ground-based and space-based telescopes. There are two general types of solutions. One tries to minimize the light of the star by directly blocking it inside the telescope, while the other tries to minimize the light of the star by combining it in at least two telescopes and eliminating it through interference. The latter device is known as an interferometer, the former as a coronograph. The most ambitious interferometer proposed is a flotilla of telescopes orbiting around the Sun and maintaining a precise formation. The design is often associated with NASA's Terrestrial Planet Finder project and the European Space Agency's (ESA) Darwin project. Webster Cash, in "Detection of Earth-like Planets Around Nearby Stars Using a Petal-shaped Occulter," *Nature*, July 6, 2006, has put forward a similarly ambitious proposal for an enormous coronograph telescope in space.

4. In special cases, when the planets are young, large, and orbit far from their stars, it is possible to discover them directly, as in the spectacular infrared images of star HR 8799 with its coterie of four planets found by Christian Marois et al., "Direct Imaging of Multiple Planets Orbiting the Star HR 8799," *Science* 322 (2008): 1348; and Marois et al., "Images of a Fourth Planet Orbiting HR 8799," *Nature*, December 23, 2010.

5. SIM PlanetQuest was a NASA mission that was in a detailed design phase a few years ago. S. Unwin et al., "Taking the Measure of the Universe: Precision Astrometry with SIM PlanetQuest" (Astronomical Society of the Pacific, January 2008).

6. A. Wolszczan and D. Frail, "A Planetary System Around the Millisecond Pulsar PSR1257 + 12," *Nature* 355 (1992): 145.

7. Pulsars are the remnants of supernova explosions—the end product of the development of a star about ten times more massive than our Sun. Even if the original star had planets, the planets around the remnant pulsar today are not those. We do not have a good idea how the pulsar planets formed after the explosion of the star, and what these planets are made of is not clear.

8. The technique was proposed by M. Holman and N. Murray, "The Use of Transit Timing to Detect Terrestrial-Mass Extrasolar Planets," *Science* 307 (2005): 1288, and by E. Agol et al., "On Detecting Terrestrial Planets with Timing of Giant Planet Transits," *Monthly Notices of the Royal Astronomical Society* 359 (2005): 567, with the practical use of the transit method in mind—transit timing variations. However, discovering an unseen planet by watching its effect on the orbit of a known planet has a venerable history. This is how the planet Neptune was discovered.

9. J. Lissauer et al., "A Closely Packed System of Low-Mass, Low-Density Planets Transiting Kepler-11," *Nature* 470 (2011): 53. In the case of Kepler-11, all planets were discovered by the transiting method, but transit timing variations allowed for the candidate planets to be confirmed and their masses measured.

10. When applied to stars, the effect is technically referred to as gravitational microlensing, in order to distinguish it from lensing between galaxies.

11. A. Einstein, "Lens-like Action of a Star by the Deviation of Light in the Gravitational Field," *Science* 84 (1936): 506; S. Mao and B. Paczynski, "Gravitational Microlensing by Double Stars and Planetary Systems," *Astrophysical Journal* 374 (1991): L37.

12. J. P. Beaulieu et al., "Discovery of a Cool Planet of 5.5 Earth Mass via Microlensing," *Nature*, January 26, 2006; D. Overbye, "Astronomers Briefly Glimpse an Earth-like Planet," *New York Times*, January 25, 2006.

13. OGLE stands for the Optical Gravitational Lensing Experiment, a US-Polish project that uses a telescope in Chile to detect stellar gravitational lensing events.

14. As stars orbit the center of the Milky Way Galaxy and we observe them from our own orbit in the Galaxy, they all appear to shift, albeit very slowly, with respect to each other. Occasionally they will literally pass in front of each other from our point of view. This is the moment when for a brief period of time we can see the gravitational bending of light—the effect of gravitational lensing. The smaller the mass of the lens, the briefer the event. With typical orbital speeds of stars in our Galaxy (200–300 km/sec) and our own motion in the same general direction, the typical duration of a stellar lensing event is several weeks. The lensed star appears to brighten, peak, and then fade back to its original brightness; the peaks typically last just a few days. The signature of a planet is a separate peak—a blip that is superposed on the brightening of the star and lasts for less than a day. Under rare favorable circumstances the orbital motion of the planet may be discernable.

15. A. Gould et al., "Microlens OGLE-2005-BLG-169 Implies That Cool Neptune-like Planets Are Common," *Astrophysical Journal* 644 (2006): L37.

## Chapter Four

1. The march, called "The Transit of Venus March," was written by John Philip Sousa in 1882 for the nineteenth-century transit of Venus and to honor the first secretary of the Smithsonian Institution.

2. The transits of Venus occur either in pairs separated by an eight-year interval or as a single transit every 121 years; three transits in a short sequence never occur. We live in an era when the transits of Venus come in pairs. The present era started with the transit in 1631 and will end with the transit in 2984, followed by a cycle of single transits. The mechanics of this are nicely described by Eli Maor in *Venus in Transit* (Princeton: Princeton University Press, 2004).

3. W. Sheehan and J. Westfall, *The Transits of Venus* (New York: Prometheus, 2004).

4. See the amazingly successful community of amateur astronomers observing extrasolar planet transits on www.transitsearch.org and the American Association of Variable Star Observers. In 2007 the former succeeded in discovering the transits of an extrasolar planet that had been discovered by the Doppler shift method—HD 17156b.

5. Stars orbiting each other, as well as spinning on their axis, are randomly oriented in the Galaxy, as studied for many decades in orbits of binary stars.

6. The argument is purely geometrical: if the distribution of inclinations is random, then the probability of transit is ($R_s/2a$).

7. The Doppler shift method discovery was described in a paper by T. Mazeh et al., "The Spectroscopic Orbit of the Planetary Companion Transiting HD 209458," *Astrophysical Journal* 352 (2000): L55, while the photometric detection of the transit was announced in the circulars of the International Astronomical Union by David Charbonneau et al. (IAU Circular 7315, 1999) and G. Henry et al. (IAU Circular 7307, 1999).

8. Based on the single known transiting planet HD 209458b, with its fairly deep transits and very "quiet" star, and a simple extrapolation ignoring many subtleties of the simultaneous photometric measurement of many stars (rather than a single one, moreover—with a known phase of the planet's orbit), estimates of the number of transiting planets that would be detected within a year run into the hundreds!

9. A. Udalski et al., "The Optical Gravitational Lensing Experiment. Search for Planetary and Low-Luminosity Object Transits in the Galactic Disk. Results of 2001 Campaign," *Acta Astronomica* 52 (2002): 1 and supplement on page 115.

10. The star is the bright star Beta Persei, an eclipsing binary star named by the Arab astronomer Al Gul (a.k.a. Algol), meaning "The Ghul's Head" and also referred to as the "Devil's Star," most likely because Arab astronomers noticed its regular "blinks." Italian astronomer Geminiano Montanari (1633–1687) noted its variability in 1670. John Goodricke (1764–1786), a celebrated English astronomer in the field of variable stars, rediscovered its variability, determined that it is strictly regular, and understood the nature of the dimming as eclipses.

11. We described the procedure, as it is now generally applied to all transit searches, in a series of papers. G. Torres, M. Konacki, D. Sasselov, and S. Jha, "Testing Blend Scenarios for Extrasolar Transiting Planet Candidates. I. OGLE-TR-33: A False Positive," *Astrophysical Journal* 614 (2004): 979; and "New Data and Improved Parameters for the Extrasolar Transiting Planet OGLE-TR-56b," *Astrophysical Journal* 609 (2004): 1071.

12. G. Torres et al., "Testing Blend Scenarios," 979.

13. M. Konacki, G. Torres, S. Jha, and D. Sasselov, "An Extrasolar Planet That Transits the Disk of Its Parent Star," *Nature* 421 (2003): 507.

14. An excellent account of this and many other stories can be found in *The Taste of Conquest* by Michael Krondl (New York:

Ballantine, 2007), a well-researched and entertaining account of one of the main motivations behind the European age of exploration—the spice trade.

15. "CCD" stands for "charged coupled device," a silicon chip with 10-micron-size pixels arranged in rows and columns, detecting light and registering its brightness as an electrical charge at each pixel.

16. One station is the Fred Lawrence Whipple Observatory (FLWO) of the Smithsonian Astrophysical Observatory (SAO) on Mount Hopkins in Arizona with four telescopes, and the other is the rooftop of the Submillimeter Array Hangar (SMA) of SAO atop Mauna Kea, Hawaii. These telescopes are modest 0.11m diameter f/1.8 focal ratio telephoto lenses that use front-illuminated CCDs at five-minute integration times.

17. A transit lasts a couple of hours and recurs every few days (for a hot Jupiter), so it is easy to miss them if you have gaps in coverage. Even worse, you often end up with many partial transits (starts or ends) that can be completely useless because the photometry during start and end of night often has systematic errors.

18. Our proposal to NASA did not go through, but the drive to discover and study super-Earths grew stronger. The term "super-Earth" (as well as "super-Venus") seems to have appeared first in print in our NASA proposal—we called the imaging/spectroscopic mission ESPI; Melnick et al., "The Extra-Solar Planet Imager (ESPI): A Proposed MIDEX Mission," *Bulletin of the American Astronomical Society* 34 (2001). The context was spectral differences, namely, that we could distinguish between a gas giant, an ice giant, and a super-Earth spectrophotometrically in reflected light. Our ESPI team did not pay attention to refining the criteria for what we called a super-Earth, as the case was marginal for any detection anyway. It was a convenient shorthand and sounded better than "fat-Earth," a short-lived suggestion in an email written by Tim Brown. In the Kepler proposal, which was written at about the same time as

ESPI, we never gave a name to the 2 and 10 Earth-mass planets. I joined the NASA Kepler team in 2000, a NASA mission that was powerful enough to deliver super-Earths and real analogs to Earth.

The term was used again in Valencia, O'Connell, and Sasselov, "Internal Structure of Massive Terrestrial Planets," *Icarus* 181 (2006): 545, a paper we wrote in 2004, though this time I made an effort in defining it—in terms of mass (1–10 Earth-mass). For the first super-Earth to be discovered (GJ 876d) by Rivera et al. ("A 7.5 M Planet Orbiting the Nearby Star, GJ 876," *Astrophysical Journal* 634 [2005]: 625), the authors did not use a specific term. The discoveries to follow were all based on the Doppler method and hence mass became the defining parameter as the term "super-Earth" was adopted by the observers.

We had an open discussion on the topic at a workshop in Nantes in June 2008. O. Grasset and I pushed for my view—to call all RV-detected planets below 10 Earth-mass "super-Earths" for the time being, since we'd be unable to distinguish between the subclasses of rocky super-Earths, ocean planets, and mini-Neptunes, and sort things out later. There were all kinds of opinions. For example, Michel Mayor and S. Udry suggested we limit the lower mass bound at 2 Earth-mass. Some were suggesting higher upper mass limits (up to 20–30), and some French colleagues were suggesting alternative names. In the end, there was no particular consensus.

19. Sunspots are temporary perturbations of the solar photosphere, the gaseous shiny surface of the sun, caused by the complex tangles of the solar magnetic field near the sun's surface. A sunspot is a region of the photosphere that has a lower temperature than its surroundings and hence appears black against the shiny surface of the sun. Sunspots are often the size of the projected circle of Mercury or bigger and can change little over days and even weeks. They appear to move slowly as the sun rotates, which is in the same direction as the planets, and in almost the same plane.

20. Eli Maor, *Venus in Transit* (Princeton: Princeton University Press, 2004), gives a detailed account of Gassendi's life and transit observations.

21. See Maor, *Venus in Transit*.

22. Maor, *Venus in Transit*, 27.

23. K. Chang, "Puzzling Puffy Planet, Less Dense Than Cork, Is Discovered," *New York Times*, September 15, 2006.

24. See Maor, *Venus in Transit*.

## Chapter Five

1. Peter Ward and Donald Brownlee, *Rare Earth* (New York: Copernicus, 2000); see a more general view of the Universe as a whole in Paul Davies, *The Cosmic Jackpot: Why Our Universe Is Just Right for Life* (New York: Orion, 2007).

2. I've been asked about this choice of name—super-Earth. The story goes back to 1999–2000, when I helped write an innovative proposal to NASA for a planet-finding space telescope with a square-shaped mirror. My colleagues Costas Papaliolios and Peter Nisenson had devised this unusual design to minimize stellar glare and allow glimpses of planets huddled close to their stars. Led by our experienced space missions scientist Gary Melnick, our team prepared a detailed scientific and engineering proposal. My job was to figure out what kind of planets our telescope might be able to discover. It seemed that planets smaller than Neptune (i.e., very large versions of Earth and Venus) were within its reach. I liked to call them super-Earths and super-Venuses for short, as it has been common in astronomy to use the adjective "super" for newly discovered or hypothesized objects that are larger in size or energy than known ones. The shorthand ended up in our publication, Melnick et al., "The Extra-Solar Planet Imager (ESPI): A Proposed MIDEX Mission," *Bulletin of the American Astronomical Society* 34 (2001): 559. It is now widely used.

3. The first super-Earth was discovered by E. Rivera et al. in 2005 and followed up by J. P. Beaulieu et al. in the same year. Many more followed.

4. The Kepler mission measures only radius (a planet's mass could be determined by separate observations in some cases), so our team has adopted a radius-based nomenclature currently. We call planets "super-Earth-size" when their radius is less than 2.0 Earths but larger than 1.25 Earths. The upper limit of 2.0 corresponds to a 10 Earth-mass planet with no bulk water and nominal range of Fe/Si ratios, similar to Earth.

5. E. Rivera et al., "A 7.5 M Planet Orbiting the Nearby Star, GJ 876," *Astrophysical Journal* 634 (2005). The name of the star is simply the consecutive number (876) in a catalog of nearby stars compiled by Gliese in 1969.

6. S. Udry et al., "The HARPS Search for Southern Extra-Solar Planets. XI. Super-Earths (5 and 8 M) in a 3-planet System," *Astronomy and Astrophysics* 469 (2007): 43. The Gliese 581 planetary system consists of a hot Neptune of at least 25 $M_E$ in a very short orbit discovered by the same team: Bonfils et al., "The HARPS Search for Southern Extra-Solar Planets. VI. A Neptune-Mass Planet Around the Nearby M Dwarf Gl 581," *Astronomy and Astrophysics* 443 (2005): 15. The two super-Earths are farther out. Unfortunately no transits were seen, so we do not know their size or exact mass.

7. D. Valencia, R. O'Connell, and D. Sasselov, "Internal Structure of Massive Terrestrial Planets," *Icarus* 181 (2006): 545; D. Valencia, D. Sasselov, and R. O'Connell, "Radius and Structure Models of the First Super-Earth Planet," *Astrophysical Journal* 656 (2007): 545.

8. D. Valencia, D. Sasselov, and R. O'Connell, "Detailed Models of Super-Earths: How Well Can We Infer Bulk Properties?" *Astrophysical Journal* 665 (2007): 1413.

9. Common materials like honey and peanut butter are very viscous. The latter is about 100 times more viscous; amorphous

solids like glass are another 10 times more viscous, but still far below the 1018 times more viscous mantle.

10. "Convection" is the physics term for a large-scale motion of fluid or gas up and down in a gravitational field due to a heat source and density differences. An example is air thermals that rise due to the Sun heating the ground, especially in the summer. The hot air near the surface rises up and is replaced by colder air coming down, and so on.

11. The notation for mantle perovskite $(Mg,Fe)SiO_3$ means that the mineral is a mixture of $MgSiO_3$ and $FeSiO_3$; for example, $(Mg_{0.6},Fe_{0.4})SiO_3$ means that 60 percent is in the former, and 40 percent is in the latter.

12. Ice VII is a cubic crystal with two interpenetrating lattices; it has a mean density of 1.65 grams per cubic centimeter (g/cc) at room temperature and exists at pressures higher than 2.5 GPa. Ice X forms after further compression of Ice VII and is denser at 2.5 g/cc; in Ice X the hydrogens are equally spaced between the oxygens. M. Choukroun and O. Grasset, "Thermodynamic Model for Water and High-Pressure Ices up to 2.2 GPa and down to the Metastable Domain," *Journal of Chemical Physics* 127 (2007): 124506.

13. For super-Earths that are relatively young and large, the temperature in the interior might reach thousands of degrees. Under such high temperature water might be in an even more exotic form known as superionic water phase: the oxygens are still "frozen" in place, just as in ice VII and X, but the hydrogens (protons) can move around.

14. Water is common in the Universe because both hydrogen and oxygen are very abundant.

15. For early work on ocean planets, see M. Kuchner, "Volatile-rich Earth-Mass Planets in the Habitable Zone," *Astrophysical Journal* 596 (2003): 105; A. Leger et al., "A New Family of Planets? Ocean-Planets," *Icarus* 169 (2004): 499.

16. Steven D. Jacobsen and Suzan Van Den Lee, *Earth's Deep Water Cycle* (American Geophysical Union, 2006).

17. See Valencia, O'Connell, and Sasselov, "Internal Structure of Massive Terrestrial Planets."

18. See L. Elkins-Tanton and S. Seager, "Coreless Terrestrial Exoplanets," *Astrophysical Journal* 688 (2008): 628; Valencia, Sasselov, and O'Connell, "Detailed Models of Super-Earths": 1413.

19. Carbon will be mostly in the form of carbon monoxide gas, CO, and largely inaccessible to newly forming planets, while the excess oxygen will be in water, silicates, and other oxides with different metals. In any case, all the silicon will end up bonding with oxygen, not carbon.

20. E. Gaidos in 2000 and M. Kuchner in 2005 described the properties of carbon planets. Overabundance of carbon in a planetary system can be inferred from analysis of the spectrum of the parent star, but such stars are extremely rare and not "normal" in many ways.

21. The planet Gliese 436b was discovered by Butler et al., "A Neptune-Mass Planet Orbiting the Nearby M Dwarf GJ 436," *Astrophysical Journal* 617 (2004): 580, using Doppler shifts. In 2007 the team of Gillon et al., "Detection of Transits of the Nearby Hot Neptune GJ 436b," *Astronomy and Astrophysics* 472 (2007): 13, found that the planet is actually transiting its parent star, which allowed the determination of its size. A careful study by G. Torres refined the mass and radius of Gliese 436b, derived by Butler et al., "A Neptune-Mass Planet." From its mean density it appears to be a Neptune-like planet, yet a very hot one, orbiting its star just seven stellar radii away every three days.

22. Computed near-infrared spectra of mini-Neptunes are markedly different from those of super-Earths due to the very different pressure scale heights in a hydrogen envelope/atmosphere: E. Miller-Ricci, S. Seager, and D. Sasselov, "The Atmospheric Signatures of Super-Earths: How to Distinguish Between Hydrogen-Rich and Hydrogen-Poor Atmospheres," *Astrophysical Journal* 690 (2009): 1056.

23. See simulations by R. Marcus, D. Sasselov, L. Hernquist, and S. Stewart, "Minimum Radii of Super-Earths: Constraints from Giant Impacts," *Astrophysical Journal* 712 (2010): 73.

## Chapter Six

1. Some amusing stories, reported originally by the *Times Berlin* correspondent, appeared in the *New York Times* on April 4, 1874.

2. John Sinkakas, ed., *Humboldt's Travels in Siberia, 1837–1842: The Gemstones by Gustav Rose* (Tucson, AZ: Geoscience Press, 1994).

3. A regular octahedron consists of eight equilateral triangles; it looks like two pyramids connected at their bases. An octahedron has six vertices. In perovskite there is an oxygen atom in each vertex, and the octahedrons are seen as connected at each vertex in each direction; the Si atom is in the middle of the octahedron.

4. Superconductivity is a curious (and very useful) phenomenon courtesy of quantum physics. A superconductor is a material that conducts electricity with zero resistance. Superconductivity was discovered and commonly occurs at the lowest of low temperature, near absolute zero. The discovery by Muller and Bednorz in 1986 was a real breakthrough because it showed superconductivity at 36 K. That is still terribly cold by human standards, but very high compared to the near 0 K of yore.

5. *National Audubon Society Field Guide to Rocks and Minerals* (New York: Knopf, 2000).

6. Post-perovskite is a high pressure phase of perovskite $MgSiO_3$, discovered by M. Murakami et al., "Post-Perovskite Phase Transition in $MgSiO_3$," *Science*, May 7, 2004; and A. Oganov and S. Ono, "Theoretical and Experimental Evidence for a Post-Perovskite Phase of MgSiO3 in Earth's D Layer," *Nature* 430 (2004): 445.

7. The nanoscale corresponds to scales/distances measured in nanometers (10–9 m) and is typical of the distances between atoms

in small molecules and crystals. By this token, "nano" has become a prefix used commonly for fabricated structures at that scale (e.g., nanolayers, nanowires, etc.), as well as for nanotechnology itself.

8. See D. Sasselov, D. Valencia, and R. O'Connell, "Massive Terrestrial Planets (Super-Earths): Detailed Physics of Their Interiors," *Physica Scripta* 130 (2008): 14035.

9. Thermonuclear energy is the source of energy in the Sun and involves the transmutation of hydrogen into helium with no radioactive by-products. It is not to be confused with nuclear energy, which involves radioactive decay to unstable heavy elements, like uranium.

10. T. Mashimo et al., "Transition to Virtually Incompressible Oxide Phase at a Shock Pressure of 120 GPa: Gd3Ga5O12," *Physical Review Letters* 96 (2006): 105504.

## Chapter Seven

1. William Shakespeare, *As You Like It*, 2.7.

2. The impending collision between the Andromeda galaxy and the Milky Way is described by T. J. Cox and A. Loeb, "The Collision Between the Milky Way and *Andromeda*," *Monthly Notices of the Royal Astronomical Society* 386 (2007): 461. Currently the velocity of Andromeda is not known with enough accuracy for scientists to definitively predict the collision.

3. Erwin Schroedinger, *What Is Life?* (Cambridge: Cambridge University Press, 1944).

4. The quantum scale was discovered and explored in the first half of the twentieth century, when quantum mechanics—the part of physics that deals with the phenomena at this scale—was developed. The word "quantum" stands for the indivisible unit of energy, a concept that was introduced in order to explain the behavior of atoms, their electrons, their interaction with light, and the emission of light. Even at very low temperatures particles at the quantum

scale (small molecules, atoms, electrons) are in constant motion and interaction with each other and with units of light (photons).

5. William H. Press, "Man's Size in Terms of Fundamental Constants," *American Journal of Physics* 48 (1980): 597.

## Chapter Eight

1. Tzvetan Todorov, *Nous et les autres* (Paris: Editions du Seuil, 1989).

2. It is probably prudent to avoid the concept of a definition when it comes to life. We do not understand life as a phenomenon well enough to define it convincingly. The unity of biochemistry of all known life on Earth means that any definition will be based on a single example, and it will be very difficult to identify what features are essential. C. Cleland discusses the issue of defining life in *Geology* 29 (2001): 987. Others would argue that no clear threshold is crossed between inert and living matter, and hence life is not yet a precise scientific concept (see "Meanings of Life," *Nature*, June 28, 2007, 447). An attempt to define life is described in a well researched and insightful report by the National Research Council, *The Limits of Organic Life in Planetary Systems* (Washington, DC: National Academies Press, 2007); D. Deamer, "Special Collection of Essays: What Is Life?" *Astrobiology* 10 (2010): 1001; George Whitesides, lecture before the Nobel Symposium on Origins of Life, 2006. For my purposes, a list of essential attributes is sufficient.

3. More precisely, an ordered network of chemical reactions.

4. More precisely, an energy-dissipating, out-of-equilibrium system.

5. To be more complete, life is adaptive, self-optimizing, fed back, forward, and stable to perturbations.

6. H. Moravec, *Mind Children: The Future of Robot and Human Intelligence* (Cambridge: Harvard University Press, 1988). See F. Dyson, *A Many-Colored Glass* (Charlottesville: University of

Virginia Press, 2007); R. Kurzweil, *The Singularity Is Near: When Humans Transcend Biology* (New York: Viking, 2005); S. J. Dick, "Cultural Evolution, the Postbiological Universe, and SETI," *International Journal of Astrobiology* 2 (2003): 65.

7. G. Joyce et al., in *Origins of Life: The Central Concepts*, ed. D. Deamer and G. Fleischaker (Boston: Jones & Bartlett, 1994). Joyce, as well as Jack Szostak of Harvard and David Bartel of MIT, pioneered the understanding and practical application of Darwinian evolution at the molecular level: molecules capable of self-catalyzing their own replication.

8. Martin Nowak, *Evolutionary Dynamics* (Cambridge: Harvard University Press, 2006).

9. This is especially true when lateral gene transfer and symbiosis are added to the paradigm. Lateral gene transfer, also known as horizontal gene transfer, is the sharing of genes between unrelated species. Ancient lineages of microbes show evidence for such sharing (Carl Woese, "A New Biology for a New Century," *Microbiology and Molecular Biology Reviews*, June 2004); lateral gene transfer and endosymbiosis seem to have been critical for creating complex genomes in the distant past.

10. J. Baross et al., *The Limits of Organic Life in Planetary Systems* (Washington, DC: National Academies Press, 2007).

11. The cosmic microwave background radiation that permeates the entire observable Universe is today at 2.7 K and sets a common lower bound for most of the gas and dust in the vast spaces between stars and galaxies. This radiation cools with time, so it was hotter, but not by much, a few billion years in the past.

## Chapter Nine

1. The HMS *Challenger* expedition is considered to have opened a new field—oceanography. It mapped the minerals of the ocean floor, discovered a large number of new species, studied

global currents, and so on. The Apollo 17 lunar module and the second space shuttle were both named after the HMS *Challenger.*

2. *Narrative of the Cruise of* HMS *Challenger with a General Account of the Scientific Results of the Expedition by Staff-Commander T. H. Tizard, R.N.; Professor H. N. Moseley, F.R.S.; Mr. J. Y. Buchanan, M.A.; and Mr. John Murray, Ph.D.; Members of the Expedition. Partly Illustrated by Dr. J. J. Wild, Artist to the Expedition.* Parts First and Second, 1885.

3. In his very entertaining book, *Life on a Young Planet* (Princeton: Princeton University Press, 2003), my colleague Andrew Knoll lays out the different threads of evidence found in Earth's ancient rocks of microbial communities surviving, adapting, and even influencing dynamic environmental change on a global planetary scale.

4. NASA's Office of Planetary Protection (http://planetary protection.nasa.gov/pp) develops the protocols for sterilizing spacecraft before they are launched, but the methods and tolerances are only as good as our knowledge of the limits. As we discover more extremophiles, they continue to break previous records.

5. S. Basak and H. S. Ramaswamy, "Ultra High Pressure Treatment of Orange Juice: A Kinetic Study on Inactivation of Pectin Methyl Esterase," *Food Research International* 29 (1996): 601.

6. Alan T. Bull, ed., *Microbial Diversity and Bioprospecting* (Washington, DC: ASM Press, 2004), 154.

7. E. Trimarco et al., "In Situ Enrichment of a Diverse Community of Bacteria from a 4–5 Km Deep Fault Zone in South Africa," *Geomicrobiology Journal* 23 (2006): 463.

8. A. Pearson, "Who Lives in the Sea Floor?" *Nature* 454 (2008): 952–953, and references. R. John Parke did pioneering studies of the deep biosphere in the sediments and rocks below the ocean floor in the 1990s.

9. A 2009 expedition to the middle north Atlantic reports metabolically active microbes in 111-million-year-old sedimen-

tary rocks at 1,600 meters below the seabed. A. L. Mascarelli, "Geomicrobiology: Low Life," *Nature* 459 (2009): 770.

10. W. B. Whitman, D. Coleman, and W. Wiebe, "Prokaryotes: The Unseen Majority," *Proceedings of the National Academy of Sciences* 95 (1998): 6578; J. S. Lipp et al., "Significant Contribution of Archaea to Extant Biomass in Marine Subsurface Sediments," *Nature* 454 (2008): 991; A. Pearson, "Who Lives in the Sea Floor?" *Nature* 454 (2008): 952–953. Lipp and colleagues showed that most of the microbes in the sub–ocean floor sediments belong to the domain Archaea, not Bacteria.

11. For decades it was assumed that the combination of high temperature, oxygen constraints, and lack of food and energy sources would prevent any multicellular organism from surviving deep inside the crust. In 2011 a team of international researchers (Borgonie et al., "Nematoda from the Terrestrial Deep Subsurface of South Africa," *Nature* 474 [2011]: 79) discovered a nematode, *Halicephalobus mephisto* (a new species), at depths where only extremophilic microbes were known to live, surprising everyone and showing that the deep biosphere is complex.

12. See J. Annis, "An Astrophysical Explanation for the 'Great Silence,'" *Journal of the British Interplanetary Society* 52 (1999): 19; J. Scalo and C. Wheeler, "Astrophysical and Astrobiological Implications of Gamma-Ray Burst Properties," *Astrophysical Journal* 566 (2002): 723; B. Thomas et al., "Gamma-Ray Bursts and the Earth: Exploration of Atmospheric, Biological, Climatic, and Biogeochemical Effects," *Astrophysical Journal* 622 (2005): L153, regarding ozone loss.

13. A. Knoll, *Life on a Young Planet* (Princeton: Princeton University Press, 2004).

14. See J. Laskar and M. Gastineau, "Existence of Collisional Trajectories of Mercury, Mars, and Venus with the Earth," *Nature* 459 (2009): 817.

15. This is averaged from measurements over the entire planet and is very difficult to do with high precision; it is also difficult to account for all the heat precisely. See Geoffrey F. Davis, *Dynamic Earth: Plates, Plumes, and Mantle Convection* (Cambridge: Cambridge University Press, 1999). The difference between measured Earth heat loss and the theoretical estimates (which predict lower values) could be due to the abundance of radioactive elements with increasing depth, or peculiar slow motions inside the mantle. J. Labrosse and C. Joupart, "A Critical Analysis of Earth's Heat Loss and Secular Cooling," American Geophysical Union, abstract T41H-03, December 2004; Lenardic et al., "Continental Growth, the Archean Paradox, and the Global Heat Flow Paradox," American Geophysical Union, abstract V32A-01, December 2004.

16. David Stevenson, "Life Sustaining Planets in Interstellar Space?" *Nature* 400 (1999): 32.

17. Knoll, *Life on a Young Planet*.

## Chapter Ten

1. How this runaway greenhouse really works was explained thirty years ago by James Kasting, who has written a wonderful book on the subject: *How to Find a Habitable Planet* (Princeton: Princeton University Press, 2010).

2. The study of habitable zones in our Solar System and around other stars goes back to the 1960s and the pioneering work of Carl Sagan. The concept has been refined since then to involve changes of the Sun in time (M. Hart, "The Evolution of the Atmosphere of the Earth," *Icarus* 33 [1978]: 23) and to account for the response and evolution of the atmosphere (J. Kasting, "Runaway and Moist Greenhouse Atmospheres and the Evolution of Earth and Venus," *Icarus* 74 [1988]: 472). The concept of a habitable zone has been broadened to the Galaxy (e.g., Peter Ward and Donald Brownlee, *Rare Earth* [New York: Copernicus, 2000]) and beyond. Since there

are many factors that contribute to making a given planet habitable, it is best to talk about habitable potential instead. See Selsis et al., "Habitable Planets Around the Star Gliese 581?" *Astronomy and Astrophysics* 476 (2007): 1373.

3. The Lick-Carnegie team added a fourth Uranus-mass planet to the Gliese 876 system—876e, which orbits farther out at a period of 127 days. E. J. Rivera et al., "A 7.5 M Planet."

4. Not surprisingly, the super-Earth planet Gliese 581d has been the subject of detailed work trying to establish its habitability, from models of a possible atmosphere and its warming effect (most recently by Wordsworth et al., "Gliese 581d Is the First Discovered Terrestrial-mass Exoplanet in the Habitable Zone," *Astrophysical Journal* 733 [2011]: 48) to spectral signatures (L. Kaltenegger et al., "Model Spectra of the First Potentially Habitable Super-Earth—Gl581d," *Astrophysical Journal* 733 [2011]: 35). However, since none of these planets are transiting we know precious little about their size and, hence, mean density. The recently discovered planetary system Kepler 11 is a cautionary tale. Though five of the Kepler 11 planets have masses smaller than Gliese 581d, none of them is rocky or has a solid surface. They are all gas rich, like Neptune.

5. D. Valencia, R. O'Connell, and D. Sasselov, "Inevitability of Plate Tectonics on Super-Earths," *Astrophysical Journal* 670 (2007).

6. P. D. Ward and D. Brownlee devote a chapter to "The Surprising Importance of Plate Tectonics" in their book *Rare Earth: Why Complex Life Is Rare in the Universe* (New York: Copernicus, 2004). That chapter is an eloquent and detailed account of all aspects of the phenomenon as it applies to Earth, as well as to animal life.

7. Ward and Brownlee, *Rare Earth*, 203.

8. See J. Kasting, *How to Find a Habitable Planet*, for detailed descriptions of the organic carbon cycle and the inorganic carbon cycle (a.k.a. carbonate-silicate cycle) and their properties as a thermostat.

9. The estimate for the $CO_2$ cycle perturbation timescale is from Jeffrey O. Bennett et al., *The Cosmic Perspective* (Boston: Addison-Wesley, 2007). See J. C. G. Walker, P. B. Hays, and J. F. Kasting, "A Negative Feedback Mechanism for the Long-Term Stabilization of Earth's Surface Temperature," *Journal of Geophysical Research* 86 (1981): 9776–9782.

10. P. Silver and M. Behn, "Intermittent Plate Tectonics?" *Science* 319 (2008): 85.

11. Valencia, O'Connell, and Sasselov, "Inevitability of Plate Tectonics on Super-Earths," 45. The detailed theory of plate tectonics is complex and remains largely unsolved today. V. Solomatov and L.-N. Moresi, "Scaling of Time-dependent Stagnant Lid Convection: Application to Small-scale Convection on Earth and Other Terrestrial Planets," *Journal of Geophysical Research* 105 (2000): 21795; C. O'Neill and A. Lenardic, "Geological Consequences of Super-sized Earths," *Geophysical Research Letters* 34 (2007): 19204; D. Valencia and R. O'Connell, "Convection Scaling and Subduction on Earth and Super-Earths," *Earth and Planetary Science Letters* 286 (2009): 492. This is likely due to the marginal efficiency of the process on small planets like Earth and Venus. Fortunately, these details are not important to the role plate tectonics plays for the habitable potential of a planet, and makes a super-Earth particularly favorable.

## Chapter Eleven

1. In typical bacteria about 1,000 nucleotides are replicated per second. The reaction is fast because of catalysis (enzymes reducing the activation energy for the reaction), not because of kinetics.

2. A synopsis of the recordings, which are preserved in the SETI Institute archives, is available on Paul Horowitz's website: frank.harvard.edu/~paulh/unpublished/fermi.html.

3. While it was Enrico Fermi who uttered the basic question of the paradox, it was Michael Hart who provided the formal description of the issue in a 1975 publication.

4. This is the same effect that you may have heard of already: the "red shift effect" seen in all distant galaxies. Their light appears progressively redder (shifted from blue color to red color) as they are farther away. Edwin Hubble and others (including Albert Einstein) observed the effect in the early twentieth century and correctly interpreted it as an indication that the entire three-dimensional space of the Universe is expanding—every galaxy is moving away from every other galaxy. The relic photons that we call the CMB are also traveling in this same ever expanding space; hence their gradual transformation from UV and optical light to microwave and radio radiation.

5. D. Spergel et al., "Three-Year Wilkinson Microwave Anisotropy Probe (WMAP) Observations: Implications for Cosmology," *Astrophysical Journal* 170 (2007): 377, on the Wilkinson Microwave Anisotropy Probe mission. The discoverers of the CMB—E. Penzias and R. Wilson—received the 1979 Nobel Prize in physics, while the first mapping of the CMB with the space mission COBE brought the 2006 Nobel Prize in physics to George Smoot and John Mather. The new Planck mission by ESA was launched in 2009.

6. A. Loeb, "The Dark Ages of the Universe," *Scientific American*, October 16, 2006.

7. The story of the stars as the crucibles of the elements is fascinating and has been told before, most notably by Carl Sagan in *Cosmos*: "We are made of star stuff."

8. Edo Burger et al., "The ERO Host Galaxy of GRB 020127: Implications for the Metallicity of GRB Progenitors," *Astrophysical Journal* 660 (2007): 504, used GRBs at $z = 4$ to get [M/H] near 1.5, compared to DLAs, which give [M/H] $<-2.0$ but probe outskirts of

galaxies. Compare this to Sozzetti and colleagues, "A Keck HIRES Doppler Search for Planets Orbiting Metal-Poor Dwarfs. I. Testing Giant Planet Formation and Migration Scenarios," *Astrophysical Journal* 649 (2006): 428, for planet fraction versus [Fe/H].

9. F. Adams and G. Laughlin, *The Five Ages of the Universe: Inside the Physics of Eternity* (New York: Free Press, 1999).

10. Incidentally, about 9 billion years ago is also the time in the past when astronomers first begin to see a drop-off in the formation rate of stars; the rate has plummeted since then by a factor of 50 or more. See R. Bowens and G. Illingworth, "Rapid Evolution of the Most Luminous Galaxies During the First 900 Million Years," *Nature*, September 14, 2006.

11. These estimates are based on the observed high correlation between current discoveries of giant extrasolar planets and the metallicity (enrichment in heavy elements) of their parent stars. D. Fischer and J. Valenti, "The Planet-Metallicity Correlation," *Astrophysical Journal* 622 (2005): 1102. Planet formation seems to go much more efficiently once the metallicity reaches at least a tenth of that in the Solar System. Continued searches for planets in old environments poor in heavy elements (the globular cluster 47 Tuc, halo stars) have failed to find planets. Gilliland et al., "A Lack of Planets in 47 Tucanae from a Hubble Space Telescope Search," *Astrophysical Journal* 545 (2000): 47; Sozzetti et al., "A Keck." Evidence relevant to rocky planets will emerge from the Kepler mission, when the mission tallies its findings and statistical analysis, in a few years. For the time being, preliminary results summarized in papers by W. Borucki et al., "Characteristics of Planetary Candidates Observed by Kepler. II. Analysis of the First Four Months of Data," *Astrophysical Journal* 736 (2011): 19, and A. Howard et al., "Planet Occurrence Within 0.25 AU of Solar-type Stars from Kepler," *Astrophysical Journal*, preprint, ArXiv: 1103.2541 (2011), show that the trend with stellar metallicity is

still there, though it appears less pronounced than the trend for giant hot Jupiters.

12. The literature on the Fermi paradox is vast, but Paul Davies offers a thorough and very thoughtful discussion in his excellent new book *The Eerie Silence* (New York: Houghton Mifflin Harcourt, 2010).

13. The estimate was introduced in Bennett et al., *The Cosmic Perspective* (Boston: Addison-Wesley, 2007).

14. Our Kepler team made this estimate in a paper (Howard et al., "Planet Occurrence"), but only for candidate super-Earth-size planets discovered by Kepler close to their stars (within 0.25 Astronomical Unit).

15. Madau et al., "The Star Formation History of Field Galaxies," *Astrophysical Journal* 498 (1998): 106, plotted the star formation rate versus red shift.

## Chapter Twelve

1. H. J. Melosh, "Exchange of Meteorites (and Life?) Between Stellar Systems," *Astrobiology* 3 (2003): 207, explores the issue in detail and concludes that interstellar trips of meteorites are very unlikely, while exchange between planets in the same system is common.

2. For example, A. Foster and G. Church, "Towards Synthesis of a Minimal Cell," *Molecular Systems Biology* 2 (2006): 1, describe the "recipe" for making a living cell ab initio. A press release by the J. C. Venter Institute dated January 24, 2008, describes the completion of the full synthetic genome of a microbe provisionally named *Mycoplasma genitalium* JCVI-1.0; published in D. G. Gibson et al., "Complete Chemical Synthesis, Assembly, and Cloning of a M. *genitalium* Genome," *Science* 319 (2008): 1215. This was a major step that eventually led the same team to the

creation of a bacterial cell controlled by a chemically synthesized genome—*Mycoplasma mycoides* JCVI-syn1.0 (Gibson et al., "Creation of a Bacterial Cell Controlled by a Chemically Synthesized Genome," *Science* 329 [2010]: 53). See also P. Berry, "Life from Scratch," *Science News*, January 12, 2008.

3. Pier Luigi Luisi, "The Synthetic Approach in Biology: Epistemological Notes for Synthetic Biology," in *Chemical Synthetic Biology*, ed. P. L. Luisi and C. Chiarabelli (Hoboken, NJ: John Wiley, 2011), 343; John Brockman, ed., *Life: What a Concept!* (New York: Edge.org, 2008).

4. The term "synthetic biology" seems to have been introduced by W. Szybalski in 1974 in *Control of Gene Expression*, ed. A. Kohn and A. Shatkay (New York: Plenum, 1974), according to S. Benner et al., "Synthetic Biology, Tinkering Biology, and Artificial Biology: A Perspective from Chemistry," in *Chemical Synthetic Biology*, ed. P. L. Luisi and C. Chiarabelli (Hoboken, NJ: John Wiley, 2011), 69. It was reused by Barbara Hobom in "Surgery of Genes: At the Doorstep of Synthetic Biology," *Medizinische Klinik* 75 (1980): 14, and then reutilized by Eric Kool (Stanford) and others in 2000, though with somewhat different connotations. A comprehensive technical review of synthetic biology in all its different forms appears in *Nature Review's Genetics* 6 (2005): 533, "Synthetic Biology," by S. Benner and A. M. Sismour. A nontechnical review by Ed Regis appears in his nice book on the subject, *What Is Life? Investigating the Nature of Life in the Age of Synthetic Biology* (New York: Farrar, Straus & Giroux, 2008). My definition of "synthetic biology" is not the widely used one as of the time of this writing, though it is essentially the same as "chemical synthetic biology" discussed by Pier Luigi Luisi and by Steven Benner in their essays in the recent compilation *Chemical Synthetic Biology*, ed. P. L. Luisi and C. Chiarabelli (Hoboken, NJ: John Wiley, 2011). The field and its language remain largely in flux. We need a new vocabulary for the novel concepts that are being introduced with its rapid development.

5. P. Luigi Luisi, "Chemical Aspects of Synthetic Biology," *Chemistry and Biodiversity* 4 (2007): 603.

6. For a detailed, accessible discussion of minimal cell definitions and the rich history of the concept, see Regis, *What Is Life?* My colleagues Jack Szostak and George Church are working on different approaches, but they encompass the conceptual framework of the synthesis. J. Szostak, D. Bartel, and P. Luigi Luisi, "Synthesizing Life," *Nature*, January 18, 2001; and A. Forster and G. Church, "Towards Synthesis of a Minimal Cell," 1.

7. There is some tantalizing evidence that cosmic and planetary environments might influence the choice of symmetry of some biomolecules; see D. Glavin and J. Dworkin, "Enrichment of the Amino Acid L-isovaline by Aqueous Alteration on CI and CM Meteorite Parent Bodies," *Proceedings of the National Academy of Science USA* 106 (2009): 5487.

8. A mirror system might allow molecular biology experiments that suffer from less contamination and are easier to perform to high fidelity. If minimal cells can be maintained, their accelerated evolution might teach us the basics of designing a minimal genome, akin to what Jack Szostak refers to as a "protocells arms race." J. Szostak, "Learning About the Origin of Life from Efforts to Design an Artificial Cell," Konrad Bloch Lecture, Harvard University, November 23, 2010. Such a genome might be central to the transition from prebiotic chemistry to biochemistry.

9. Examples of this approach that are relevant to origins of life research are the pioneering work of the A. Eschenmoser Group, for example, A. Eschenmoser, "Searching for Nucleic Acid Alternatives," in *Chemical Synthetic Biology*, ed. P. L. Luisi and C. Chiarabelli, (Hoboken, NJ: John Wiley, 2011), 12. Breakthrough work on nucleotides synthesis in prebiotically plausible planetary conditions was done by the J. Sutherland Group. M. Powner, B. Gerland, and J. Sutherland, "Synthesis of Activated Pyrimidine Ribonucleotides in Prebiotically Plausible Conditions," *Nature*, May 14, 2009.

10. Generation II life is different from Generation I evolving (via cultural evolution) to a postbiological state, as discussed by Steven Dick, "Cultural Evolution, the Postbiological Universe, and SETI," *International Journal of Astrobiology* 2 (2003): 65, and references; or the "closer to us humans who transcend biology" of Ray Kurzweil or the Homo evolutis of Juan Enriquez.

11. The story follows the research done after the discovery of well preserved mummies in burials dating 3,000–4,000 years ago in the Tarim basin, around several ancient cities that later became an essential part of the Silk Road, as described in J. Mallory and V. Mair, *The Tarim Mummies* (London: Thames & Hudson, 2000).

12. Sven Hedin, *Der Wanderde See*, 2nd ed. (Leipzig: Brockhaus, 1941).

13. J. Mallory and V. Mair, *The Tarim Mummies*.

# INDEX